SUBTERRANEAN IMAGINARIES AND GROUNDWATER NARRATIVES

This book interrogates the problems of how and why largely unseen matter, in this case groundwater, has found limited expression in climate fiction. It explores key considerations for writing groundwater narratives in the Anthropocene.

The book investigates a unique selection of climate fiction alongside an exploration of hydrosocial environmental humanities through a focus on groundwater and groundwater narratives. Providing eco-critical analysis, with creative fiction and non-fiction excerpts interwoven throughout, and drawing on Indigenous Australian and Australian settler novels and poems alongside European, American and Japanese texts, the book illuminates the processes of 'storying with' subterranean waters – their facts, uncertainties, potencies and vulnerabilities. In a time when the water crisis in an Australian and worldwide context is escalating in response to global warming, giving voice to the complexities of groundwater extraction and pollution is vital. Drawing from non-representational, posthumanist and feminist perspectives, the book provides an important contribution to transnational, comparative climate fiction analysis, enabling an interdisciplinary exchange between hydrogeological science and the eco-humanities.

This book is an engaging read for scholars and students in creative writing, environmental humanities, cultural and post-colonial studies, Australian studies, and eco-critical literary studies. Writers and thinkers addressing the problems of the Anthropocene are called to pay attention to the importance of subterranean imaginaries and groundwater narratives.

Deborah Wardle teaches environmental literature and writing RMIT and Creative Writing at the University of Melbourne, Australia. She is a researcher, with numerous projects completed, including with the Office of Indigenous Education and Engagement at RMIT University and the 'Groundwater Narratives' project with Melbourne Sustainable Society Institute at Melbourne University. She has had stories and peer-reviewed articles published in Australian and international journals including *Meanjin*, *Overland*, *The Big Issue*, *Meniscus*, *Mosaic* (Canada), *Fusion*, *Swamphen* (ASLEC-ANZ) and *Animal Studies Journal*.

Routledge Environmental Humanities
Series editors: **Scott Slovic** (University of Idaho, USA), **Joni Adamson** (Arizona State University, USA) and **Yuki Masami** (Aoyama Gakuin University, Japan)

The *Routledge Environmental Humanities* series is an original and inspiring venture recognising that today's world agricultural and water crises, ocean pollution and resource depletion, global warming from greenhouse gases, urban sprawl, overpopulation, food insecurity and environmental justice are all *crises of culture*.

The reality of understanding and finding adaptive solutions to our present and future environmental challenges has shifted the epicenter of environmental studies away from an exclusively scientific and technological framework to one that depends on the human-focused disciplines and ideas of the humanities and allied social sciences.

We thus welcome book proposals from all humanities and social sciences disciplines for an inclusive and interdisciplinary series. We favour manuscripts aimed at an international readership and written in a lively and accessible style. The readership comprises scholars and students from the humanities and social sciences and thoughtful readers concerned about the human dimensions of environmental change.

For more information about this series, please visit: www.routledge.com/Routledge-Environmental-Humanities/book-series/REH

SUBTERRANEAN IMAGINARIES AND GROUNDWATER NARRATIVES

Deborah Wardle

Routledge
Taylor & Francis Group
LONDON AND NEW YORK

earthscan
from Routledge

First published 2024
by Routledge
4 Park Square, Milton Park, Abingdon, Oxon OX14 4RN

and by Routledge
605 Third Avenue, New York, NY 10158

Routledge is an imprint of the Taylor & Francis Group, an informa business

British Library Cataloguing-in-Publication Data
A catalogue record for this book is available from the British Library

ISBN: 978-1-032-21881-6 (hbk)
ISBN: 978-1-032-21882-3 (pbk)
ISBN: 978-1-003-27041-6 (ebk)

DOI: 10.4324/9781003270416

Typeset in Joanna
by codeMantra

The book is dedicated to Deb Shaw

CONTENTS

ARTWORKS

ABOUT THE ARTIST

Bill Sampson has a broad contemporary drawing, painting, sculpture, and, more recently, photographic and video practice. He studied at the Slade in London before returning to the University of Melbourne to gain a PhD in 2007 with a thesis titled *Prettybad: towards an aesthetics of the expressionless*. He lectured in Critical and Theoretical Studies in the School of Fine Arts and taught life drawing at the VCA.

Among other prizes and awards, Sampson's work has been exhibited at the Korean International Art Fair, has been a finalist in the National Works on Paper Prize, and has been awarded a Grant for New Work by the Australia Council. Closer to home, Sampson has been a finalist in the Arthur Guy Memorial Painting Prize, Bendigo Gallery, and the Len Fox Painting Award 2016, Castlemaine Gallery.

ACKNOWLEDGMENTS

I acknowledge the Dja Dja Wurrung, the people on whose unceded lands I have written this book. I respectfully acknowledge their Elders, past and present. I thank the Dja Dja Wurrung Cooperative (Victoria, Australian) for assistance with use of Indigenous language in the novel excerpts from *Why we cry*.

There are many people to thank for their part in the creation of this book.

Linda Daley and Cathy Greenfield were my supervisors during the original research for my PhD. Their wisdom is built into the foundations of this book.

Teaching creative writing, contemporary eco-fiction, and environmental literature to undergraduate and master's students over the past seven years has immeasurably inspired my ideas for this book. I appreciate greatly the inspiration from colleagues and students in my ongoing teaching and learning.

Following my PhD, the seeding grant project 'Understanding Aquifers through Groundwater Narratives' through the Melbourne Sustainable Society Institute (MSSI) at University of Melbourne enabled me to further develop my research into groundwater narratives. My work in this project was wonderfully supported by hydrogeologists Peter Dahlhaus and Tim Peterson, and creative writing scholar and artist Amanda Johnson. Support from Brendan Gleeson, Sangeetha Chandra-Shekaran and John Wiseman from MSSI enabled me to continue with the work.

Early readers of this manuscript are angels in disguise. I thank the anonymous reviewers of the proposal for this book, whose feedback and suggestions helped move the manuscript from 'dissertationese' to something hopefully more mature. Manuscript edits by Kate Gleeson greatly improved the text. Feedback from Dr Amanda Johnson and hydrogeologist Heather Vick on the draft manuscript of this book was invaluable.

Early readers of the draft fictional manuscript during my PhD years include Heather Vick (hydrogeologist), Peter Dahlhaus (hydrogeologist), Bruce Lindsay (water lawyer), Jeanine Leane Indigenous scholar and poet), Sue Bruce, and my son Oliver Wardle. Their insightful contributions are greatly appreciated.

I am a reading group junkie and have benefited greatly from discussions and explorations of readings for many years. This has included: the Higher Education Learning Program (HELP) group from the first years of my doctorate at RMIT; EcoFeminist Fridays through the University of Melbourne; the newly formed Affect Theory Readers Group; and the Posthuman Reading group (members from RMIT and beyond). Colleagues in the Nonfiction Lab – RMIT, have supported my work and thinking throughout. Also at RMIT, School of Global Urban and Social Studies staff and students, where I taught subjects in Climate Change Responses and Communicating Sustainability, reinforced my expressions of the wide impacts of climate change and the importance of environmental justice, in particular Ben Cooke and Nooshin Torabi from the Sustainability and Urban Planning team.

I thank the editor from *Meanjin* for permission to include an excerpt from 'Springs Sojourn' which first appears in *Meanjin*, 82(2), Winter 2023. I also thank publisher Magabala Books for permission to include the complete poem 'Prologue and Response' by John Kinsella and Charmaine Papertalk-Green from their collection of poems, *False claims of colonial thieves* (2018).

The images at the start of each chapter are by Dr Bill Sampson, whose intelligent application of his work to each chapter greatly enhances the book. His images illustrate how coalescence between surface and subterranean water and imaginary places can expand and shape imaginary invocations of groundwater's presence.

I thank all those from the Routledge – Environmental Humanities Series who have steered this book to completion, in particular Matthew Shobbrook, whose clear communications have guided and supported me throughout the process.

Most significantly I thank my life partner and great love Deb Shaw. Her consistent presence and support nourish my internal life, enabling the subterranean work of writing to continue.

1

THE WATER CYCLE

INTRODUCTION

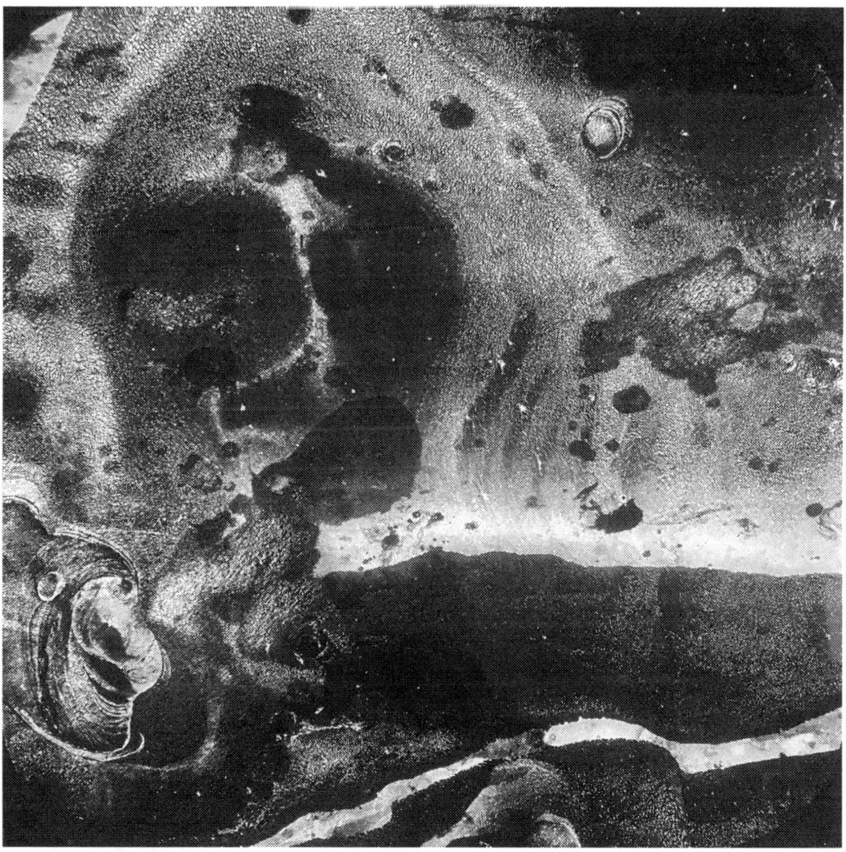

Figure 1.1 Bill Sampson, from *Anomalies were no longer news* (1 of 4), 2020, sump oil and synthetic polymer on paper, 22 × 22 cm.

DOI: 10.4324/9781003270416-1

The water cycle is commonly understood as the global and localised movements of water as it condenses from atmospheric vapours and clouds into rain, some of which meets surface waters (rivers, lakes and oceans). Surface water evaporates, to again join the atmospheric waters. The water cycle describes the movements of water over time and across places. Some water, usually less than 5 per cent, seeps into aquifers. Ageless limits to water on planet Earth, that is the same amount of water cycling through eons, are affected by human and geographical factors. Groundwater's role in the water cycle is only recently becoming appreciated. Groundwater's place in the hydrosocial cycle, a wider concept that includes human impacts on water movements, is not well understood.

> This is where the giant serpent continues to live deep down under the ground in a vast network of limestone aquifers.
>
> Wright 2006, p. 2

This book explores the ways that climate writers can engage with and express groundwater's seeming invisibility, its potency and its contemporary vulnerability. Numerous suggestions and provocations are offered throughout the chapters to prompt climate writers and thinkers to consider the stories that groundwater can tell. Subterranean imaginaries ask writers and scholars to make an 'inward migration', to reach internally to unlock a source of writing and thinking *with* groundwater's material and metaphoric potency (Wright 2022). As demonstrated in the epigraph above from Wright's (2006) novel *Carpentaria*, this book invites readers to enter the world of aquifers, to give attention to subterranean imaginaries.

Groundwater is a rare player in Australian literature, under-represented in fiction, poetry and creative non-fiction. Hydrogeology is a small but vital field of scientific study of groundwater composition and flows. Aquifers, as divergent and inscrutable subterranean water bodies, reveal themselves, not only as organically rich bodies of affective flow, but as sites that spark imaginations and bring underground watery matter to new significance.

The book explores concepts for storying *with* groundwater. I introduce and conceptualise groundwater as narrative groundwater. The concept *narrative groundwater*, or *storied groundwater*, establishes the capacity of groundwater to narrate, to express its own forces and forms, through imaginaries that enmesh the materiality of subterranean waters *with* writers', poets' and scholars' words. Narrative groundwater uncovers the material and textual reality and relationality of groundwater in its many storied iterations. Conversely, I call the storied renditions of narrative groundwater, *groundwater narratives*. I propose that widening understandings of groundwater are well served by processes of storying and creating imaginative narratives based on sustainable relations between human and more-than-human entities. Storying is a creative practice method to develop diverse writings and artistic expressions that give voice, agency and relationality to subjects, in this case to groundwater. Humanities scholars partaking in the hydrological turn have thought of surface waters in many ways – metaphorical water as conceptualised by Bachelard (in Chen *et al.* 2013, p. 10), modern water, global water (Linton 2010), gestational water, bodies of water (Neimanis 2017),

Anthropocene water, to name only some. The concept of narrative groundwater builds on the water scholarship of Linton (2010) and Neimanis (2017), drawing attention to subterranean imaginaries to recognise the vital place that groundwater plays in the global water cycle and unfathomable human and nonhuman lives.

This book explores how climate writers and creative writing scholars can give narrative expression to groundwater. By storying *with* groundwater, I draw attention to the social, cultural and political imperative to engage with the 'storied matter' of nonhuman actors (Iovino and Oppermann 2014). Writing groundwater narratives, I will argue, enables enmeshments between human and nonhuman ecologies to be affirmed and performed through a 'storying *with*' writing process (Wardle 2019). The various ways that climate change discourse influences how humans make meaning of our world and of ourselves are still relatively unexplored in literature (Baucom and Omelsky 2017; Schneider-Mayerson *et al.* 2020; Hennessy *et al.* 2022). Literary contributions towards meaning making in the Anthropocene through fiction, poetry and creative non-fiction featuring groundwater continue to be partial and emerging. Climate writing, as will be discussed throughout the book, continues to provide significant inroads to understanding of human and more-than-human relationality.

Water is transformative in biological, sociological, political, and economic terms. Water supply is ever changing, more critically with the ongoing double risks of scarcity and inundation, and questionable water economics (Rockström *et al.* 2023). However, Jamie Linton argues:

> defining the water crisis in terms of diminishing quantities of available water obfuscates the bigger picture – namely that every issue involving water is realized in a specific social and cultural context. ... water scarcity is the product of social and economic circumstances. ... So long as we accept uncritically the notion that water is about water, we acquiesce to fixing its nature in a particular web of language, knowledge, law, management, and material infrastructure that does not serve social purposes very well in either water-rich or water-poor regions.
>
> Linton 2010, p. 227

Cross-disciplinary research contributes to the growing scholarship in the environmental humanities field.[1] Interdisciplinary water discourse, a previously small sub-discourse of climate change studies, is emerging as a driving force behind expressions of water's life-giving and deadly actions. Groundwater sits quietly, at times invisibly, in interdisciplinary considerations of water's roles and potency. It is not just that groundwater is water. It is the qualities and affects that groundwater assumes by being underground, slow-moving and almost invisible that fuel its expressions in the following chapters and creative writing examples. Groundwater's affective voice focuses our attention on its perambulating flows through permeable geologies and provides an analogy for the ways that human emotions are expressed. I discuss the literal and metaphorical affect of groundwater's expressions in more detail in Chapter 6. Utilising literary imagination as a vehicle to broaden understandings and to engender changed relationships with subterranean worlds is a significant challenge within anthropocentric responses to climate change. Through a convergence of creative writing, social and cultural theory, groundwater science and ecocritical literature, I braid groundwater's contemporary urgent plight into diverse climate writing.

Understanding groundwater's scale and its place in the water cycle is a useful foundation. As mentioned above, the water cycle describes the global, regional and local movements of water in its various forms as it condenses from air into clouds and water droplets, and falls as rain. Some water is absorbed into soils and subsoils to be accessible to plants. A smaller proportion seeps more deeply into aquifers to flow beneath regions and continents in subterranean groundwater basins or aquifers. Some water trickles and moves across landscapes in creeks, streams and rivers towards oceans. Water transpires from forests and grasslands and evaporates from lakes, dams and oceans to once again form atmospheric waters and clouds. The cycle continues endlessly at varying rates of transformation over time and in different places.

Groundwater's influence in the global water cycle is only recently being researched. Ninety-seven per cent of the world's water is salty, existing in the seas and oceans. The remaining 3 per cent is fresh water, found in the icecaps, in some aquifers, lakes, rivers, and in the atmosphere as clouds. Groundwater comprises approximately 96 per cent of the accessible freshwater reserves across the planet (Ladson 2008). Groundwater provides a significant proportion of the world's fresh water for domestic, agricultural and industrial use. Not all groundwater is potable, or 'fresh'. At varying depths and ages, groundwater absorbs salts and minerals from the surrounding rocks and substrates. Water's time underground may range from minutes to years, or to centuries, or in some deep, slow aquifers for thousands of years. Old groundwater may be very salty. Only in recent years has the scientific hydrological community begun to understand the range of connections between global groundwater reserves and surface water, and groundwater's role in the broader actions of the surface water cycle (Leap 1999; Ladson 2008). Groundwater, like clouds, does not follow international or state borders, nor property boundaries. Analysis of conflicts and dilemmas in both Australian and international scenarios are an important element of groundwater's storying potential (Jarvis 2014).

The term hydrosocial cycle reflects the interjections of human endeavours and their influences on water's movements (Linton and Budds 2014). It is in many ways more pertinent than simply considering the hydrology of the water cycle. The hydrosocial cycle 'represents the process by which alterations or manipulation of water flows and quality affect social relations and structure, which in turn affect further alteration or manipulation of water' (Linton and Budds 2014, p. 175). 'Water is what we make of it' is the opening line of Jamie Linton's (2010) study of the genealogy of 'modern water' – the water that has come to be seen as a resource to be managed and commercialised for human usage. Narrative groundwater accumulates knowledges similarly to modern water, building cultural and social interpretations of the waters beneath our feet.

Many areas of the world are facing increasing threats of water crises, either increased flooding events or catastrophic long-term droughts. In considering local and global water crises, following Linton (2010), this book considers

> how modern water itself has created the conditions for its own crisis and how, instead of a water crisis, what we are facing is the crisis of modern water. Framing the problem in this way suggests that we need to change the way we think about water.
>
> Linton 2010, p. 192

Groundwater's potency and fragility in the Anthropocene, its scale and invisibility, its links to ecological and anthropogenic calamities, and the fact that in its subterranean milieu it cannot be directly experienced in the manner of flood, storm and tempest, puts it in need of new modes of narration.

At either a continental level or at the smaller scale of landscape catchments, water balance is a term that reflects the changes in water quantities held in storage (e.g., lakes, rivers, aquifers) due to changes in inputs and outputs (Ladson 2008, pp. 5–6). One of the effects of climate change has been escalating droughts. As the lengths and severity of droughts increase and surface water storages are only intermittently replenished, groundwater is increasingly used as a water source (Ladson 2008; Brodie *et al.* 2012). The limits to finite groundwater supplies across the planet are increasingly being tested as human boring for groundwater now often exceeds the capacity of aquifers to recharge. Such scenarios provide critical storying possibilities.

Dry conditions, long droughts and intermittent flooding are normal features of Australian weather patterns. For example, the ten-year drought in Australia at the start of the twenty-first century ended with torrential floods in 2010–11. Known as the Millennial Drought, it prompted the Australian government to revise the national water laws in efforts to consolidate access to potable water. The legal labyrinth of Australian water law has done little to remedy the effects of changing climate patterns on water's accessibility. I briefly mention the complex arena of legal, social and political dimensions of groundwater management and governance, and water law, a field of investigation of its own. Groundwater management plans, which are meant to ensure the sustainability of aquifers, are often based on old data or suffer from no historic data at all (Michael *et al.* 2011; Jarvis 2014; Nelson *et al.* 2020). Around the world, groundwater's usage is often unregulated by laws of the land. Australian water law and groundwater licensing systems are the envy of many countries. Water licensing laws vary from state to state in Australia, causing dissent across borders where the same aquifers are pumped according to different laws. For example, the limestone aquifer which lies under the Victorian and South Australian borders is managed by the Border Groundwaters Agreement – South Australia–Victoria (1985 and amended in 2005), but has not prevented ongoing conflicting demands from agriculture, industrial and domestic use in each state.

In Australia the problems of groundwater extraction or pollution are numerous. These include the damage to the Barwon Downs aquifers by pumping from the extensive bore field to supply water to Geelong; the mound springs in Central Australia, which are unique desert spring-fed ecosystems at risk of extinction due to agriculture and over-extraction of groundwater from the Great Artesian Basin; receding aquifers that supply townships in the Murray-Darling Basin; pollution to the Great Artesian Basin through coal-seam gas fracking in the Pilliga Forest (north-central New South Wales); open-cut coal mining on the Liverpool Plains, NSW; contamination of aquifers beneath and surrounding numerous airports where seepage of PFAS chemicals in fire retardants has polluted groundwater supplies for miles around; excessive use of superphosphates in some agricultural areas; and seepage of urban wastewater which pollutes groundwater reserves in numerous locations. There are many other examples. This book suggests thinking and writing processes that highlight the science behind ecological and public safety impacts for communities using groundwater. Toxic narratives abound revealing despoliation of specific groundwater sites around the world. For example: land subsidence and sinkholes where aquifers are excessively pumped causing damage to buildings and infrastructure in Maceio, Brazil, Mexico City and many cities; drying step wells from drawdown caused by deep commercial bores in India limiting community access to water; and saltwater intrusion to aquifers under coastal cities due to over-pumping freshwater aquifers, affecting building foundations (Brooklyn, New York and Sydney, Australia are examples). Accessible groundwater narratives give voice to these and many other occurrences across global and local scenarios.

Groundwater sustains innumerable human and nonhuman lives. Access to groundwater in Australia and in many settings worldwide is often a source of conflict between those who want unrestricted rights to a scarce resource, often for financial rewards, and those who seek preservation of groundwater reserves for wider community, cultural and environmental use. Learning of the different values ascribed by various groups to groundwater's existence and usage opens the possibility of understanding the variability and complexity of narrative groundwater (Gibbs 2006; Jarvis 2014). Commercially based extraction of groundwater often creates conflicts with human communities and nonhuman ecologies seeking sustainable survival. Storying *with* specific groundwater contexts is a writing process explored in more detail in Chapter 3. Groundwater narratives uncover the conflicts between competing groundwater users, rich fodder for climate writers.

Climate change refers to long-term meteorological patterns across global scales, usually utilising at least thirty years of data. Day-to-day or short-term, locally experienced weather events fit within the broader impacts of climate change (Neimanis and Walker 2020). Weather events are sometimes confused with climate change, as the intra-actional relationship between weather and longer-term climate patterns is constantly changing. It has been argued that global warming is a diluting term for the multiple effects of climate change, one that dislocates the impacts of capitalism and global inequities from catastrophic environmental impacts (Baucom and Omelsky 2017). Climate crisis or climate catastrophe are alternative terms. As climate change persistently pushes weather patterns into new, unexpected extremes, groundwater too is being forced to the brink, pumped to its limits. Regular assessment reports from the Intergovernmental Panel on Climate Change (IPCC) show the worsening effects of climate change over the past fifty years. The 2022 report to the United Nations was particularly damning, indicating that global atmospheric carbon levels continue to increase, and the planetary warming consequences of the greenhouse effect have continued to escalate since the reports commenced in 1988 (IPCC 2022). In the context of climate change, narrative groundwater renders stories across both short and long timeframes and across a myriad of settings. Groundwater narratives straddle the immediacy of human and nonhuman relations and the shifting timeframes of groundwater's actions.

Climate Writing

Anthropocene fiction, climate fiction, or the shorthand term *cli-fi*, is broadly defined as a literary genre that tells stories of the causes and effects of climate change and global warming (Trexler 2015; Andersen 2020).[2] Climate fiction that engages with accurate depictions of science or imagines speculative futures is well appraised by Adam Trexler in *Anthropocene fictions: the novel in a time of climate change* (2015) and more recently by Gregers Andersen in *Climate fiction and cultural analysis: a new perspective on life in the Anthropocene* (2020). Groundwater, however, is a rare literary entrant. I contend that the influences of groundwater on landscapes and human existences are as pervasive as the imagination can construct.

As the science of global greenhouse gas accumulations leading to warming atmospheres and oceans became better understood, climate change began to enter cultural imaginaries more stridently in the 1990s. Amitav Ghosh (2016) raised the difficulties in narrating the creeping impacts of climate change, occurring over large timeframes, calling the absence of such stories a failure of imagination, 'the great derangement' of contemporary times. Michael Segal (2017) similarly bemoaned the missing climate change stories, citing the lack of direct 'experience' of climate change as an explanation. Finding ways to understand and tell stories about groundwater's role in global environmental change entails wide-scale scientific and political thought that delves beneath surface perceptions of global environmental changes.

Despite Amitav Ghosh's fears of 'the great derangement', or a lack of literary expressions of climate change, many recent scholars now point to the burgeoning body of fiction and non-fiction writing that engages with the impacts of global warming (Trexler 2015; Fetherston 2016; Ghosh 2016; Bradley 2017; Brooker 2018; Johns-Putra 2018; Sussex 2018; Andersen 2020; Schneider-Mayerson *et al.* 2020; Cole 2022). Growing threats of extreme weather events, leading to either or both severe surface water scarcity and/or intermittent flooding are now well represented in climate writing, particularly in the fast-growing field of climate fiction (Trexler 2015; Ritson 2019; Andersen 2020; Hennessy *et al.* 2022). It has been argued that any contemporary fiction is by necessity climate fiction (Fetherston 2016). Climate fiction has become a fast-growing genre globally. Adam Trexler argues that 'To date nearly all Anthropocene fiction addresses the historical tension between the existence of catastrophic global warming and the failed obligation to act. … [It] offer[s] a medium to explain, predict, implore, and lament' (2015, p. 9). I add the possibility that climate fiction also has a responsibility to inspire hope and agency in its readers. I will throughout this book provide examples of narratives that reveal hope and positive actions for change alongside the irrefutable catastrophes and dilemmas of living through climate change. The book explores how writers might consider giving subterranean water, sometimes known as bore or well water, a voice at this critical time.

Telling stories of underground places evokes unbounded potential for metaphors and analogies, a writer's terrain. The paucity of representation of groundwater in Australian literatures is here referred to as *gaps in the water-literature cycle*. Groundwater's uncommon presence in many Western societies' cultural expressions of water creates literary potential that is yet to be plumbed by climate writers aiming to increase awareness and understanding of groundwater's narrative capacity (Richard-Ferroudji and Lassaube 2020, p. 192). Narrating the influences of subterranean places and groundwater's permeable, soaking journeys invites climate writers into processes of thinking, storying, integrating the sciences of hydrogeology with groundwater's affective and agential capacities, to enable the voices of subterranean imaginaries to be expressed.

Subterranean in the *Macquarie dictionary* is an adjective meaning: '1. existing, situated, or operating below the surface of the earth; underground; 2. existing or operating out of sight or secretly; hidden or secret' (Macquarie Dictionary 2010). The concept of an underworld, often fiery or hellish, sometimes hollow and habitable, has been imagined by writers for centuries. Dante's *Divine comedy* starts with the section 'Inferno', where Dante accompanies Virgil on a journey to hell to discover the meaning of sin and evil's consequences. Many seventeenth- through to twentieth-century writers have explored the idea of subterranean worlds in science and speculative fictions. Peter Fitting (2004) anthologises the main players in the underground literature realm in *Subterranean worlds: a critical anthology*. Fitting's descriptions and extracts from many subterranean imaginings from 1665 to 1914 starts with descriptions of early beliefs in a hollow earth. Fitting thoroughly covers the lengthy and almost forgotten tradition of writing set in fantastical underworlds. Jules Verne's classic science fiction *Journey to the centre of the Earth*, first published in 1864 in French, remains a classic with many translations. Verne's subterranean adventurers leave readers with the message that science is the key way for humans to uncover truth. Stories of buried treasures also invoke images of humans digging to underground spaces. As I will explore in later chapters, the aim of this book is not to expand on the body of literature that speculates on human journeys to the underworld; rather, I limit this exploration to considerations of how subterranean imaginaries can build depictions of groundwater's responses to the Anthropocene. Writers and storytellers have focused on the earth's surface responses to climate change. Ways of looking beneath the surface, particularly ways to foreground how subterranean waters sustain healthy water cultures, is the central concern here.

To *imagine* subterranean water flows, to create subterranean imaginaries in text, calls for a rudimentary discussion of imagination and ways of imagining. Again starting with the *Macquarie dictionary* definition of the verb, to imagine means:

> 1. to form a mental image of (something not actually present to the senses). 2. to think, believe, or fancy. 3. to assume or suppose. 4. to

> conjecture or guess: *I cannot imagine what you mean.* 5. *Obsolete* to plan, scheme, or plot. *–verb (i)* 6. to use the imagination. 7. to suppose; think; conjecture.
>
> Macquarie Dictionary 2010

The noun 'imagination' can mean the action or faculty of imagining, or of forming mental images or concepts of what is not actually present to the senses (Macquarie Dictionary 2010). Reproductive or creative imagination recombines memories and experiences into new images. To imagine may also mean producing 'ideal creations consistent with reality, as in literature', and is distinguished from 'fancy' or capricious and desultory fantasies (Macquarie Dictionary 2010). Literary imagination invokes all these possibilities. In this book I use the term *imaginaries* to include both the noun and the verb senses of the word. Imaginings could also be used. Imaginaries implies the outputs of imaginations as well as the processes of imagining the unseen. Subterranean imaginaries indicate the ways that writers and thinkers play with creative and philosophical ideas to imagine groundwater's realms.

The book explores storying as a writing method to imagine and narrate the many human and nonhuman perspectives embedded in groundwater stories. It explores how climate writers might write stories of places that we cannot directly see. Subterranean imaginaries invite us into an unseen underground, into watery aquifers. Thinking alongside the materiality of groundwater, as Karen Barad suggests, enables writers to experiment with 'new imagining in its materiality' (2015, p. 388). The imaginings become material as writers craft groundwater narratives. This endeavour has an underlying politics of expressing vital groundwater narratives. The imaginaries have real impacts. Subterranean imaginaries are well expressed by Barad:

> Imaginaries with material existences in the thick now of the present – imaginaries that are attuned to the condensations of past and future condensed into each moment: imaginaries that entail superpositions of many beings and times, multiple im/possibilities that coexist and are iteratively intra-actively reconfigured: imaginaries that are material explorations of the mutual indeterminacies of being and time.
>
> Barad 2015, p. 388

Following this thread, imaginaries of groundwater are not controlled by an author, but by writerly or authorial sensitivity to the conglomeration of groundwater expressions as these move through past, present and into future timeframes. Groundwater can begin as the seeping of rain into subsurface soils, that then sinks and soaks into saturated geological zones. It can take anything from minutes to days to months for water to reach aquifers, and it then flows according to gravity's pull and geological conditions for anything from days to centuries before it reaches the surface again. No longer groundwater, it becomes a familiar surface body of spring water, a swamp, a lake. Groundwater's multi-timed travels can be imagined by listening to histories and observing ecosystem responses, by learning through hydrogeology, and by responding to the affects that groundwater produces.

Throughout the book I use the term *climate writers*, or *climate writing*. This is to imply firstly that I am considering the ideas and processes of writing groundwater narratives in the context of anthropogenic climate change and its myriad impacts, predominantly in the twenty-first century. Climate writing links the global with the local. Writers seeking to consider subterranean imaginaries as a means to enrich climate change writing and scholarship will, I trust, find this book useful. Second, the generic term *climate writing* refers to a range of possible forms and styles of writing. Writers ask questions about 'whether or not the form we are adopting for our purposes is the best kind of vehicle to do the job about what it is we want to say' (Rendell-Short 2015, p. 93). The presence, the grandeur and the intricate enmeshments of the minuscule in environments are frequently described through fiction, creative non-fiction, poetry, the lyric essay, film scripts and documentaries, to name a few generic possibilities. Each form is open to the reciprocity between text and ideas, between ecologies of human and nonhuman lives, that storying *with* invites. There is no one form of creative climate writing to explore human expressions of groundwater's presence and influence, and its vulnerability to the incursions of humans. As a fiction writer, with a special interest in climate fiction (cli-fi), the discussion in this book focuses on the fictional form. The philosophical ideas and frameworks discussed throughout the book are however applicable to many forms of climate writing.

In the cyclic, conjunctional ways of the water/hydrosocial cycle, the book explores literary exemplars to see how climate writers have given meaning and voice to this marginal, seemingly silent entity, this underground place. I select specific works of fiction and several poems to illustrate the role of literature in making groundwater visible in the hydrosocial cycle. These include Alexis Wright's *Carpentaria* (2006), Marcel Pagnol's *The water of the hills* (1962), Elizabeth Jolley's *The well* (1986), W.H. Auden's poem 'In praise of limestone' (1948) and Banjo Paterson's poem 'Song of the artesian waters' (1896). A writer reads to improve their own work (Brien 2006; Boulter 2007; Prose 2012). This book presents literatures that invoke ways to understand groundwater, so as to enhance climate writers' efforts to engage with the challenges of climate change.

Adam Trexler (2015, pp. 223–225) discusses how Anthropocene fiction has tended to avoid realist depictions of climate change, more often being written as speculative, futurist, dystopian or science fiction. Dystopian climate writing of apocalyptic futures has prevailed in climate fictions and is all too quickly becoming current realities. The 'slippage between realism and dystopia' is increasingly blurred in contemporary climate writing that describes the horrors of the real impacts of climate change (Mills 2022, p. 41). Through what Mills calls 'dystopia creep' (2022, p. 39), speculative, dystopian or science fiction writing genres may perhaps defer reader's connection to the immediacy of climate change dilemmas. A greater problem with dystopian climate-related stories is the propensity of bleak or horrific depictions of climate change to generate fears that paralyse readers, reducing their capacity to engage in positive actions to address climate change (Cole 2022; Hennessy *et al.* 2022; Mills 2022). In their discussion of how 'spec-fic' (an abbreviation for a wide range of speculative fiction forms) responds to contemporary political and environmental challenges, Cat Sparks and Rose Michael argue that 'the movement between the "real" – however that might be defined – and the speculative is a source of great energy and inspiration' (Sparks and Michael 2018, p. 2). Groundwater narratives may have elements of conjecture, creating 'energy and inspiration' in depictions of real problems and solutions. Importantly, speculations about hopeful futures in groundwater narratives aim to acknowledge human and nonhuman agency in responses to climate change. Each of the possible literary styles for interrogating climate change have their distinct characteristics and relevance. Each will appeal to various spectrums of writers and readers who bring distinct questions to their research/writing/reading.

Climate fiction, sometimes referred to as Anthropocene fiction (Trexler 2015), is one of the fastest growing genres of literature worldwide (Schneider-Mayerson *et al.* 2020). I use the term Anthropocene throughout the book cautiously to indicate the proposed geologic epoch where human activities are acknowledged to be impacting the planet's geology, climate and ecological systems. I acknowledge that the term is contested and has difficulties. The International Union of Geological Sciences has not yet approved the term as an official depiction of a new geological time scale. Humanity's responsibility for generating new global climatic and geologic patterns is however widely recognised. In the humanities the Anthropocene has become a much-debated term, adopted and critiqued in may disciplines and forums. Donna Haraway's (2016) injunction to 'stay with the trouble' and write stories that counter the anthro- of the Anthropocene is significant in the context of this book. Making places for stories in what she calls the Chthulucene (cthulhu is from the Greek *khthon*, which means earth and retains the agency of the Earth to determine its own systems) enables a more coherent merging of human and more-than-human reciprocities (Haraway 2016). Other problems with the term Anthropocene include the question who do we mean by the 'Anthro'? Which humans? The term is gender and racially blind and takes no account of the role of capitalism, industrialisation or corporatisation as significant systemic factors behind global geological markers (Cohen *et al.* 2016, pp. 7–8). Assuming even particular humans' contributions to global warming, such as burning fossil fuels and detonating atomic bombs, lets the systemic structures that are causing global warming off the hook. After decades of ignoring the science and being deluded by opportunistic climate deniers, innovative measures to mitigate and adapt to climate change are emerging. Widespread evidence for anthropogenic impacts on climate implies humanity's accountability for globally felt impacts of climate change. It does not take account of the fact that people living in poverty in the Global South, who produce the least carbon emissions, will feel the devastations of climate change the most. The generalisation of the 'Anthro' remains problematic. Despite these problems, I will use the term Anthropocene for its capacity to ask humans to think into geologic timeframes of the past and the future, as many creative writers are endeavouring towards (Johns-Putra 2018, p. 40 fn 1). Dipesh Chakrabarty, an influential thinker about the term Anthropocene, proposes in his essay 'The climate of history: four theses' that the term Anthropocene acknowledges humans as geologic agents, not only biologic agents, changing our imaginations of what it means to be human (Chakrabarty 2009, p. 206). The Anthropocene may be an inadequate term, but as a concept for climate writers to remind climate readers of human responsibility, it is a starting point.

The book draws attention to the thinking behind the writing strategies of climate writers in an age when we are rapidly approaching an ecocide tipping point (Chakrabarty 2009; Cohen *et al.* 2016; Hamilton 2017). Creative writers and scholars across the environmental humanities are invited to consider the possibilities of relational thinking and storying *with* groundwater, as a way forward in bleak times.

Outline of Theoretical Scope

This book crosses broad, interdisciplinary theoretical terrain. Giving attention to matter and mattering, the book invites writers and scholars to consider the vital materiality of human and nonhuman intra-relationality as primary. For too long divisions and oppositions between humans and nonhuman animals and entities have resulted in cruelty and damages. Environmental destruction and species extinctions have proliferated under capitalist assumptions that certain humans can control and use the resources of the planet without propriety or taking responsibility for sustainability. Theorists and thinkers who have influenced the new materialist approach of this book include Jane Bennett (2010), Gilles Deleuze and Felix Guattari (1988), Rosie Braidotti (2013) and Donna Haraway (2016). Many other thinkers have also argued that seeing and writing *with* the subjectivity of matter or things at the forefront is an important ethic to writing in the Anthropocene (Massumi 2010; Ingold 2011; Chen *et al.* 2013; Iovino and Oppermann 2014; Neimanis 2017). Understanding that meaning can be derived from material things through storying is foundational to my endeavours.

In thinking about groundwater narratives, Astrida Neimanis's concept of 'bodies of water' locates water within and beyond human bodies:

For us humans, the flow and flush of waters sustain our own bodies, but also connect them to other

> bodies, to other worlds beyond our human selves. Indeed, bodies of water undo the idea that bodies are necessarily or only human. The bodies from which we siphon and into which we pour ourselves are certainly other human bodies (a kissable lover, a blood transfused stranger, a nursing infant), but they are just as likely a sea, a cistern, an underground reservoir of once-was-rain. Our watery relations within

> (or more accurately: *as*) a more-than-human hydrocommons thus present a challenge to anthropocentrism, and the privileging of the human as the sole or primary site of embodiment.
>
> Neimanis 2017, p. 2

Neimanis draws on Adrienne Rich's politics of location to construct relations between human bodies and places. Bodies of water, for Neimanis, encapsulate both human and more-than-human bodies and are deeply influenced by place. Neimanis's posthumanist, feminist stance on embodiment provides a point of access to writing groundwater narratives that refer to deep bodies of water, invisible bodies of water, waters that intermingle with bodily waters, waters contained within and beyond human stories (Neimanis 2017). Groundwater narratives build on Neimanis's re-figuring of the embodiment of human bodies of water as always intermingled/enmeshed with more-than-human bodies of water (Neimanis 2017, p. 67). Such imaginings of groundwater narratives reveal how human relations with a watery world might be re-figured. I will show that imagining groundwater narratives that put aside mastery and profit in place of deeper co-constituted respect for hydro-common materialities becomes the aim of groundwater narratives. Considering groundwater through this lens opens multiple ways of writing its presence, its affects, its places and performances in wider ecological, agricultural and industrial environments that entangle humans in watery webs.

I write of groundwater's multiple performances as a material and metaphoric entity. Groundwater's vulnerability in response to climate change is increasingly being studied (Novicky *et al.* 2010). Groundwater's susceptibility to anthropogenic misuse indicates the importance of exploring an aesthetic portrayal of this otherwise unseen player in global human and environmental enmeshments. The book explores how climate writers might not only know about, but also create, narrative meaning from aquifers in a relational, globally warming world.

A posthuman-materialist perspective in writing aims to reduce the centrality of human positioning. This may at first glance appear to imperil the use of human language. Can climate writing be a legitimate method to impute human responsibility for climate change, and to demonstrate a material relationally with nonhuman animals and entities that can enliven new thinking? I cautiously navigate the tensions of being a human writer-scholar, utilising language to emphasise the possibilities of a posthumanist performativity. Barad (2008, p. 122) argues that we need to shift away from mimetic efforts of representation towards discursive practices that blur and merge the social and the scientific. Such an approach, I will argue, is useful to writers considering expressions of groundwater in climate writing.

The book draws attention to how writers and scholars come to know of 'a life' in the material watery entities they are storying *with* (Bennett 2010). Bennett follows the ideas of Gilles Deleuze in suggesting the 'indeterminate vitality' of materials traditionally seen to be inanimate (Bennett 2010, pp. 53–55). This book proposes how writers might consider the politics of representing 'matter' in ways that enable the performances of its more-than-human characters to be understood as lively and agential. These interlinked precepts are plaited into an examination of methods and methodologies of climate writing and scholarship.

Jane Bennett encourages us to acknowledge the limits of a social constructionist viewpoint, whereby 'things' are understood according to the social context of their existence. Like Deleuze and Guattari, she attributes 'a life' to matter that goes beyond the social constructs viewpoint (Bennett 2010, p. 61). The long list of my social and cultural trappings and the influencing factors that seep into the research behind this book is important but is in some measure a diversion to groundwater's positioning. Being female, a mother, middle class, white, educated, lesbian, Australian, feminist, sister, friend, with innumerable life experiences, profoundly affects my relations with groundwater and with literature. Such factors contribute to the substrate that underlies my writing in the early twenty-first century. However it is groundwater's own agency, its own 'interstitial field of non-personal, ahuman forces, flows, tendencies, and trajectories' that this book explores (Bennett 2010, p. 61). I acknowledge the fragmentary nature of my humanity, and the mutual and relational connections I live amidst. I am, by my human presence, part of the 'unmaking', which is how Deborah Bird Rose (2013) describes the fragmentation of late modernist thinking, writing and acting. Writing in the Anthropocene calls for listening and thinking slowly and ethically in ways that witness and expose certainties that may be flawed (Rigby 2009; Rose 2014). Writers and scholars may acknowledge the underlying of social and political influences and the aspirational perspectives embedded in creative and scholarly endeavours. In addition, enmeshments with the material voices of matter in the Anthropocene drive scholarship to new places. Here, the watery spheres of subterranean aquifers envelope the material domain encountered.

This book articulates relationships between creative writing processes and posthuman thought. Groundwater, as the subject of my inquiry, is the kind of material entity that has historically eluded direct representation in literary discourse. I eclectically synthesise the sciences of groundwater with philosophy, and with the arts and environmental humanities, in ways that cultivate understandings of the more-than-human, the more-than-textual, through events, practices and assemblages of everyday possibilities. Non-representational viewpoints emphasise 'the fleeting, viscous, lively, embodied, material, more-than-human, precognitive, non-discursive dimensions of spatially and temporally complex lifeworlds' (Vannini 2015, p. 318). Vannini's non-representational approach suits the fluidity and immensity of groundwater. Under this broad umbrella, I propose an understanding of groundwater that asserts that all entities and processes, including human beings, are made of matter, along with material and physical forces (Barad 2008; Bennett 2010; Barrett and Bolt 2013). My considerations of groundwater as pluming, subterranean, moving, cycling, fluid matter, and its complicity with human lives, draws upon Karen Barad's (2017) ideas about the politics of mattering. She argues that 'material-discursive phenomena are constituted through each other, each in specifically entangled ways. … The very stuff of the world is a matter of politics' (Barad 2017, p. G117). Macarena Gomez-Barris's (2017) aspirations also apply well to my thinking in relation to the extraction of groundwater:

> I explicitly challenge the frames of disciplinary knowledge that would bury the subtlety and complexity of the life force of the worlds that lie within the extractive zone. … Through detailed studies of local spaces, Indigenous worlds, and natural ecologies in peril, I work to better conceptualize histories of conquest and dissent …
>
> Gomez-Barris 2017, pp. xv–xvi

The complexities of groundwater mining – where groundwater is extracted at rates known to be beyond the recharge capacity of aquifers – exemplify the influence, the ethics and politics attached to groundwater. Also, the effects of groundwater pollution – recognising that how we deal with industrial and agricultural wastes at the surface has implications for groundwater into the far distant future. Fuel spills, use of toxic fire retardants and many other industrial pollutants will affect groundwater by seeping into aquifers for perhaps centuries. I aim for readers to move past fear and anger, and to be moved to consider current threats to ecologically sustainable groundwater flows. To entreat readers to be moved, to encourage care about the sustainability of groundwater are important aspirations of this book.

The book finally invites fresh considerations as to how climate writing merges, enmeshes and diffracts usually disparate bodies of knowledge. I write towards an interplay of the science of hydrogeology, new materialism, posthuman perspectives, the influences of affect and the imaginings of climate writing. In these interdisciplinary contributions to creative writing research the theoretical considerations are not so much an 'appendage' to the creative work, as the bedrock to creative writing practice (Krauth 2011). Excerpts from various fictional and poetic narratives seep through the ground of the theoretical thinking, telling a story of groundwater, illustrating the coherence of thinking and writing with subterranean imaginaries. Groundwater's narrative 'textuality' reveals how groundwater can be read, and how it becomes 'storied matter' (Iovino and Oppermann 2014, pp. 6–7).

Consistent with posthuman questioning of the pre-eminence of 'mankind', feminist perspectives also influence my approach to writing groundwater narratives. I have been inspired by many eco-feminists who challenge masculine assumptions of dominance and have explored the conundrums of women's relationships to the natural world. Many eco-feminists have critiqued patriarchal, capitalist conceptions of extractive, resource-to-be-taken viewpoints that have contributed to habitat loss, species extinctions, global warming and climate catastrophes. Val Plumwood (1993), a globally recognised Australian philosopher, was an early feminist critic of patriarchal mastery of nature perspectives. Stacey Alaimo (2012) and Elizabeth Grosz (2017) are go-to feminists who developed important thinking about the trans-corporeal nature of matter and the ethics of an incorporeal conception of matter. Deborah Bird Rose (2013), an anthropologist, inspires climate epistemologies that draw together anti-colonial sentiment and feminist concerns. Lara Stevens *et al.* (2018) provide an overview of feminist ecologies through an outstanding collection of eco-feminist essays, which show the interconnectedness of the contemporary fights for both women's rights and ecological sustainability. Feminist viewpoints guide me to explore the conceptual and literal violence against women and nature (Gaard *et al.* 2013). Frankie Pankhurst, the protagonist in *Why we cry*, not only bears the surname of a foundational suffragette and activist, she is also characterised by strong independence and an interest in social justice in her journey into water activism.

Structure of the Book

This book makes conscious contributions to the ways that climate writers give expressions to important environmental concerns, spanning six main themes:

1. how we might know and not know groundwater;
2. storying *with* as a narrative technique;
3. connecting climate writing with credible science;
4. considerations of how new nature writing emerges from non-representation of place;
5. understanding the impacts of affect in climate writing; and
6. building the capacity for climate writing to enable and express agency and activism in readers.

While the themes are all interconnected, I explore their inferences in the following chapters.

Chapter 2 explores narrative epistemologies, which expand ecocritical knowledges of groundwater. It asks how humans might know and not know the significance of matter and of groundwater's existence. Thinking *with* the waters beneath our feet includes thinking about what groundwater is, how it moves, and some of the many ways that humans and wider nonhuman ecologies interact with groundwater. The epistemological foundations of how we might understand groundwater as watery matter include an examination of broad-ranging posthuman, non-representational thinkers and material ecocritics.[3] Merging the features of climate fiction, and also poetry and non-fiction writing, with posthuman, new materialist theoretical frameworks suggests a way to address writing problems of representing large-scale, unbounded, and inter-relational matter, such as groundwater. By de-centring human voices, the book argues for a permeable exchange or, as described in Chapter 2, *oscillating absorption* between the parameters of climate writing and posthuman, new materialist perspectives.

Chapter 3 draws attention to the processes of storying *with*, anchoring the book's purpose to interrogate new forms of groundwater stories, expanding ethical representations of the effects of climate change on groundwater. Storying *with* is a process of writing that presents and explores socially, culturally and ecologically responsible narratives (Phillips and Bunda 2018). Storying *with* means creating unique collaborations to produce contemporary and ethical narratives 'that bring the unimaginable … into a familial narrative space, framed and structured by an artistic storytelling practice' (Tynan 2020, pp. 48–49). Storying *with* methods are based on a dialogic and relational ethos, not expecting to impose singular perspectives, rather developing an approach to storytelling that is sensitive to a multitude of perspectives. Storying *with* is attentive to how the narrative may affect both human and nonhuman others and implicate the author's own positionality (Mertoja 2017, p. 306). Writing ecocritical stories in the age of climate change entails acknowledging that 'practical ethics come from the stories we tell' (Admussen 2016, p. 6). Using processes of storying *with*, particularly through writing, researching and engaging with the diverse interests and capacities of human and nonhuman subjects, encourages climate writers to emphasise the agencies of human and nonhuman stakeholders. Chapter 3 discusses how writers and scholars may apply an ethical, ecocritical storying *with* methodology to previously untold accounts of groundwater's roles in human and ecological survival.

The book draws together elementary aspects of the large body of hydrogeological sciences. Chapter 4 discusses the relationship between science and literature. Calculations of groundwater's flows using recharge rates of the past to make predictions about future groundwater reserves could be problematic when neither the transience of aquifers, nor the impacts of climate change are considered (Novicky *et al.* 2010; Dahlhaus *et al.* 2014; Currell *et al.* 2016). The complex history of water governance, law and policy in Australia is well represented elsewhere (Crase 2008) and beyond the scope of the book.[4] Current groundwater licensing practices in Australia are derived from the legal and historic context of white settlers' uses of groundwater, often not connected to Indigenous knowledges, nor suited to the ephemeral nature of Australian rivers and streams which connect to aquifers. Nor is water management legislation keeping pace with the vacillating demands for water as climate change impacts subterranean waters. I predominantly use the term hydrogeology to encompass the foundational concepts of groundwater science.

Groundwater narratives give emphasis to a sense of place, watery subterranean places. Chapter 5 investigates place-based literatures and writing of more-than-human places from non-representational frameworks, including a discussion of groundwater as a hyperobject. It includes a discussion of the vital role of First Nations writers in articulating relationality with place.

In Chapter 6 I explore the affective interconnections between surface water, cellular water (animal and vegetable) and bodies of groundwater. The practical aspects of groundwater's roles in human and nonhuman endeavours evoke strong emotions, such as need, dependence, greed, awe, to name a few. Affect theory has been much touted by thinkers such as Greg Seigworth and Melissa Gregg (2010), Ben Anderson (2016), Jane Bennett (2010) and Sarah Ahmed (2014). Affect in general terms refers to the visceral, the incorporeal vital forces beyond emotion, an emergent in-between-ness with a capacity to reach the inorganic and non-living (Seigworth and Gregg 2010, pp. 1–2). Bennett argues for the impacts of affect, saying, 'moods and aesthetic sensibilities influence ethics and politics as much as do words, arguments and reason' (2010, p. 61). I propose groundwater's essential capacity to invoke affect in human readers as a means to engage them with pressing global problems of climate change, and here specifically with the impacts of water scarcity on groundwater ecologies. Creating affects and emotions in readers, such as threat, fear, passion and hope, may be a common writerly aspiration. How writers and scholars navigate the raft of emotions and affects that the writing process evokes is explored in connection to ideas about permeability.

Chapter 7 concludes the book with a discussion of activism in literature, using the analogy of groundwater-dependent ecosystems to encourage readers to listen to and observe the potency of subterranean narratives. There is not one possible conclusion, but a range of possible learnings, thinking and processes that may be drawn. The bricolage of possibilities that may emerge from each chapter remains in readers' perceptions of groundwater's narrative voices. Whether readers will be led to real-world agency or political activism remains speculative. The book builds a case for the critical responsibility borne by climate writers and scholars to elucidate the vulnerability and potency of groundwater under the effects of global warming.

This book is derived in part from a PhD dissertation I completed at RMIT University, Melbourne, Australia. This doctoral study, *Storying with groundwater: why we cry* (Wardle 2019), utilised a creative practice research methodology to develop a scholarly analysis of creative writing's contributions to climate fictions, and a creative fiction manuscript titled *Why we cry*. Based on the PhD and subsequent ongoing research, four approaches to writing are interwoven throughout this book. First, I present a discussion of theoretical and scholarly thinking that underpins the development of groundwater narratives.

Second, creative writing is presented as a creative practice method for making meaning and new knowledge. The excerpts from the creative component of my PhD manuscript, *Why we cry*, are interwoven throughout, stepping this book towards a hybrid scholarly form that blends the critical and scholarly thinking behind groundwater narratives with a creative response to how groundwater's voices may be framed. The excerpts provide examples of fiction that reflect in some part the realities of rural lives dependent on groundwater. Human characters are at first glance the inevitable focus of a contemporary realist story. At the same time, the narrative weaves the presence of groundwater throughout the lives of human and nonhuman characters in ways that draw into question the centrality of human perspectives. In the excerpts I portray various possibilities, perceptions and expressions of groundwater's effects and affects. *Why we cry* tells a story of protagonist Frankie Pankhurst's journey to becoming a community water activist. This sits alongside the reluctance of her small rural community, Cawling, to respond to threats to groundwater from a proposed property development by the antagonist, Kasha Duff. Frankie's friend Wade Keane gives glimmers of hope when, late in the plot, it is revealed that he has joined his Indigenous community's fight for recognition of its rights stemming from cultural connections to groundwater. Frankie and Wade's growing understanding of groundwater run as parallel narrative threads to the increasing presence of groundwater. *Why we cry* bridges rational concepts of groundwater's materiality with relational ways of viewing groundwater as interconnected matter. Subterranean imaginaries may imply that writers are speculating, imagining worlds beyond their own immediate experience. Bridging realistic aspects of groundwater's impacts and connections with its imaginary subterranean agencies is a core aspiration of this book. The excerpts from *Why we cry* do not appear in the chronology of the manuscript, but are selected here to suit the themes of each critical chapter. The relation between the theoretical thinking and the fictional excerpts is partially explained by the analogy of the braiding and unbraiding of a river as it spreads into several branches, then converges in new streams and finally, at its delta, meets the ocean. The water's flow continues towards the sea, though different streams are formed, at times a series of anabranches push their owns path before re-entering the main river course and resuming the river flow. This analogy describes what has been called plaiting or braiding of theoretical thinking and creative writing work (Brien 2006; Krauth 2011; McKinnon 2023). The plaiting of relevant scenes from *Why we cry* demonstrates how the formal and conceptual creative writing process has merged with theoretical and scientific considerations. The book motions towards the integration of the two forms of writing; it's claims are derived from the ways that each discipline reinforces and supports the ideas of the other.

In the third approach taken in the book, I include ecocritical analysis of literary texts that mention groundwater.

Fourth, the book presents my reflections on both the creative and theoretical writing processes that support creative writing and scholarship in the 'blue humanities'.[5] It asks that readers consider the leap from thinking to action.

At the beginning of each chapter, I introduce a hydrological term that threads analogically to the content of the chapter. For example, the concept of porosity (Chapter 5) arose when I considered the unimaginably leaky spaces that writing groundwater narratives traverses. Encapsulating leaky literary terrain through concepts of porosity and permeability adds metaphorical meaning. Each chapter is also introduced by a graphic image that aims to spark and open imaginings of the themes of the chapter. The artwork of Dr Bill Sampson invites the reader into each chapter through a non-representational imputation and imaginary engagement with the content of the ensuing chapter.

Aquitards: Limits to Storying

Underground, there are limits. Aquitards are the confining layers of non-porous rock, or sometimes clay, which restrain flows of groundwater, holding them at a certain depth, or within a bound region (Fitts 2013, p. 124). The degrees of restriction in various aquitards changes over time and over distances. For example, a basalt layer may restrict water flow, until a small fissure, or chemical reaction in the rock pores, allows minute seepages, or new flows. These are known as leaky aquitards (Fitts 2013, pp. 324–326). Aquitards are a necessary part of groundwater systems. In similar ways, writers and readers face limits in storying *with* the vulnerability and potency of aquifers. Using this analogy, my work as a scholar and a fiction writer meets boundaries and limitations, which change over time. This book is bounded within several limits.

First, the book gives attention in the main to an Australian context. The groundwater underlying Australia, the driest inhabited continent, has a long and fraught story to tell of its increasing and seemingly unnoticed fragility. Problems of over-extraction and pollution of aquifers exist worldwide, but not all sites can be canvassed here.

Limiting my explorations to an Australian setting means acknowledging Country, which includes respecting Aboriginal and Torres Strait Islander stewardship and kinship with place. The term Country carries inferences of custodial relationships in ways that the terms land or landscape do not (Tredinnick 2005, p. 8; Wright 2016). An Australian focus to the book embraces sensitivity to and recognition of the sovereignty of Indigenous Australians' relationships with groundwater narratives. In writing of connection to Indigenous Country that has never been ceded to white colonialists, attempts to 'get it right', as Anita Heiss says, may be almost paralysing for white writers (Sheldon-Collins and Heiss 2014). As a white writer in Australia, I write the limits of my relationship to place. I seek respectful relationships with Indigenous connections to Country. I inhabit a stolen continent that has never been ceded. 'Always was, always will be' is a catchcry of Australian activists seeking recognition of Indigenous land rights. Hearing and expressing the voices of narrative groundwater entails an engagement with authenticated histories of colonial encroachment and Indigenous resistance. Environmental justice will not occur without acknowledging the need for justice for Australia's First Nations. Enabling the voices of places to speak is an integral element of Indigenous culture carefully navigated in this book. These stories are being best told by First Nations writers, as explored in Chapter 5. Culturally sensitive and ethical perspectives on groundwater's usage, potency and vulnerability are an underpinning weft to the following discussions.

Groundwater narratives recognise water's place in First Nations' unique heritage. Recognition of Indigenous sovereignty, resistance and agency, and/or in any instances of using First Nations cultural and intellectual property, climate writers advisedly follow the ethical protocols provided by the

Australia Council for the Arts (Janke 2019). 'Australia's unique Indigenous artistic and cultural expression is rooted in thousands of years of heritage and continuing practice and its uniqueness has been recognised internationally' (Janke 2019, p. 1).[6]

Groundwater conflicts are of concern globally. Across the planet Indigenous custodians are calling for recognition of water rights. The predominantly Australian examples of groundwater narratives and Australian literature in this book aim to model the ethics of respect due to Indigenous custodians worldwide, and also to bring focus to First Nations' peoples' struggles for water rights.

This is not a book about the confining spaces of caving, or taking one's terrestrial body to tight underground places. Robert Macfarlane (2019) takes us to magisterial and terrifying subterranean places in his book *Underland*. As he says, 'for many reasons we tend to turn away from what lies beneath. But now more than ever we need to understand the underland' (Macfarlane 2019, p. 13). Where water saturates rock is a place that is impossible for humans to physically occupy. The task of imagining underground places and multiple timeframes comes with epistemological boundaries and may at times be restricted by circumscribing social or cultural constructs. Subterranean groundwater's seeming invisibility to everyday human experience challenges how it might be expressed within the limits of human imaginings.

This book does not address the conundrums of time in storying *with* narrative groundwater. Writing *with* groundwater's time scales unsettles the 'this, then that' approach to writing. Subterranean imaginaries bleed from deep time through the recent past, through the plethora of present catastrophes, into speculative futures. The stickiness of time in groundwater narratives is not explored in any depth in the book.

Writing in the third year of the global COVID-19 pandemic was both an opportunity to immerse myself in solitary research for this book, and at the same time it reduced my opportunities for fieldwork engagements with colleagues. Despite this limit, writing a book that encourages consideration of interiority was enhanced by expanded times of isolation and quiet reflection.

Groundwater's depletion through unsustainable extraction, and its pollution by chemical wastes, evoke its vulnerability. Groundwater moves, it rises and descends with the moon, it follows gravity, it responds to geological forces. I attempt to make meanings of groundwater's watery qualities, remaining true to its tangible features as known through contemporary science. This book progresses climate change conversations between climate writers, creative writing scholars, cultural theorists and hydrogeologists. These unique interdisciplinary exchanges foster considerations of responsibilities of writers and thinkers to the politics and practices of a progressive environmental humanities. Creative writing practices that engage readers and the wider community with a desire to address the water crisis emerge from imaginative understandings of human entanglements with largely unseen subterranean bodies of water.

My encounters and experiences of reading literatures and landscapes for signs of groundwater, and my study of the science, politics and history of groundwater extraction in Australia, all influence the ways I make meaning through diverse creative works. The book explores modes of knowledge making through the practice of creative writing through the six main themes mentioned above. It makes an incursion into the problems and considerations of climate writing, considering how writers express the voices and agencies of groundwater in the more-than-human world. The book explores literary impressions of groundwater's responses to climate change.

Notes

1 See the collection of writing in the *Routledge companion to the environmental humanities* for an overview of issues and developments in this emerging scholarly discipline, and concisely summarised in the Introduction (Heise 2017).

2 The terms climate fiction, climate writing, climate change writing and Anthropocene fiction are used interchangeably in this book. Adam Trexler chooses the term *Anthropocene fiction* to reflect human intervention in changes to the world's climate and geology (Trexler 2015, pp. 1–2).

3 For a more in-depth overview of environmental or eco-philosophy see Matthias Fritsch *et al.* (2018). In Chapter 2 I explore a selection of philosophical positionings that are relevant to thinking into writing groundwater narratives.

4 Lin Crase (2008) provides a concise summary of the challenges and changes in water law and policy in Australia.

5 The 'blue humanities' encapsulates the growing field of interdisciplinary research that addresses human and cultural relationships with marine and

surface water bodies. Groundwater is a rare player in this hydrological turn. Maude Barlow's (2013) *Blue future* provides a useful introduction to concepts of the 'blue humanities'. *Arts & Cultural Studies Review* issue 2(48) (2021) was also themed on the 'blue humanities'.

6 Janke's protocols for ethical engagement with Indigenous knowledges include in summary ten key principles:

> *Principle 1 Respect:* Giving Welcome to Country and/or Acknowledgement of Country; embracing diversity; ensuring cultural safety
> *Principle 2 Self-determination:* Enabling Indigenous leadership in projects
> *Principle 3 Communication*, consultation and consent; identifying people in authority; obtaining free, prior and informed consent
> *Principle 4 Interpretation:* Acknowledging cultural contexts and perspectives
> *Principle 5 Cultural integrity and authenticity:* accurate and authentic use of terminologies; ensuring integrity – rights of creators and communities
> *Principle 6 Secrecy and confidentiality:* Sacred and secret material must be confidential; representations of deceased people must preserve lore and personal privacy
> *Principle 7 Attribution:* Attributing communities or language groups and multiple contributors or communities accurately
> *Principle 8 Benefit sharing:* Wages and employment, other payment and conditions must be negotiated; copyright ownership in resulting works, royalties and residuals must be accurate.
> *Principle 9 Continuing cultures:* New uses of ICIP need new consents; archiving or depositing works and future custodians of ICIP must be transparent.
> *Principle 10 Recognition* and protection of First Nations rights
>
> Janke 2019

The full document provides essential reading.

References

Admussen, N., 2016. Six proposals for the reform of literature in the age of climate change. *The Critical Flame: A Journal of Literature and Culture*, 42, 1–6.

Ahmed, S., 2014. *The cultural politics of emotion*. Edinburgh: Edinburgh University Press.

Alaimo, S., 2012. States of suspension: trans-corporeality at sea. *Interdisciplinary Studies in Literature and Environment*, 9, 476–493.

Andersen, G., 2020. *Climate fiction and cultural analysis: a new perspective on life in the Anthropocene*. Abingdon: Routledge.

Anderson, B., 2016. *Encountering affects: capacities, apparatuses and conditions*. New York: Routledge.

Auden, W.H., 1948. In praise of limestone. *PoemHunter* [online]. Available from: www.poemhunter.com/poem/in-praise-of-limestone-3/ [accessed 16 August 2022].

Barad, K., 2008. Posthumanist performativity: towards an understanding of how matter comes to matter. *In:* S. Alaimo and S. Hekman, eds. *Material feminisms*. Bloomington, IN: Indiana University Press, 120–154.

Barad, K., 2015. Transmaterialities: trans*/matter/realities and queer political imaginings. *GLQ: A Journal of Lesbian and Gay Studies*, 21, 387–422.

Barad, K., 2017. No small matter: mushroom clouds, ecologies of nothingness, and strange topologies of spacetimemattering. *In:* A.L. Tsing *et al.*, eds. *Arts of living on a damaged planet: ghosts and monsters of the Anthropocene*. Minneapolis, MN: University of Minnesota Press, 103–120.

Barlow, M. 2013. *Blue future*. Toronto: House of Anansi.

Barrett, E. and Bolt, B., eds., 2013. *Carnal knowledge: towards a 'new materialism' through the arts*. London: I.B. Taurus & Co.

Baucom, I. and Omelsky, M., 2017. Knowledge in the age of climate change. *South Atlantic Quarterly*, 116, 1–18.

Bennett, J., 2010. *Vibrant matter: a political ecology of things*. Durham, NC: Duke University Press.

Boulter, A., 2007. *Writing fiction: creative and critical approaches*. Basingstoke: Palgrave Macmillan.

Bradley, J., 2017. Writing on the precipice. *Sydney Review of Books*, 21 February [online]. Available from: http://sydneyreviewofbooks.com/writing-on-the-precipice-climate-change/ [accessed 22 March 2017].

Braidotti, R., 2013. *Posthuman*. Oxford: Wiley.

Brien, D.L., 2006. Creative practice as research: a creative writing case study. *Media International Australia*, 118(1), 53–59.

Brodie, R., Lawrie, K. and Commander, P., 2012. Groundwater – lifeblood of the continent. *In:* R. Blewett, ed. *Shaping a nation: a geology of Australia*. Canberra: Geoscience Australia and ANU Press, 332–279.

Brooker, B., 2018. Climate change was so last year: writer's festivals and the great derangement. *Overland*, 3 September [online]. Available from: https://overland.org.au/2018/09/climate-change-was-so-last-year-writers-festivals-and-the-great-derangement/ [accessed 7 September 2018].

Chakrabarty, D. 2009. The climate of history: four theses. *Critical Inquiry*, 35, 197–222.

Chen, C., Macleod, J. and Neimanis, A. 2013. Introduction: towards a hydrological turn? *In:* C. Chen, J. Macleod and A. Neimanis, eds. *Thinking with water.* Montreal, QC: McGill-Queen's University Press, 3–22.

Cohen, T., Colebrook, C. and Miller, J.H., 2016. *Twilight of the Anthropocene idols.* London: Open Humanities Press.

Cole, M.B., 2022. 'At the heart of human politics': agency and responsibility in the contemporary climate novel. *Environmental Politics*, 31, 1–20.

Crase, L., 2008. *Water policy in Australia: the impact of change and uncertainty.* Washington, DC: Resources for the Future.

Currell, M., Gleeson, T. and Dahlhaus, P., 2016. A new assessment framework for transience in hydrogeological systems. *Groundwater*, 54, 4–14.

Dahlhaus, P., *et al.*, 2014. Making the invisible visible: the impacts of federating groundwater data in Victoria, Australia. *Journal of Hydroinformatics*, 18(2), 238–255.

Deleuze, G. and Guattari, F., 1988. *A thousand plateaus: capitalism and schizophrenia.* London: Athlone Press.

Fetherston, R., 2016. Greener pastures and tangled gums: the rise of Australian eco-fiction. *Overland*, 7 December [online]. Available from: https://overland.org.au/2016/12/greener-pastures-and-tangled-gums-the-rise-of-australian-eco-fiction/ [accessed 31 August 2018].

Fitting, P., 2004. *Subterranean worlds: a critical anthology.* Middleton, CT: Wesleyan University Press.

Fitts, C.R., 2013. *Groundwater science.* Waltham, MA: Academic Press.

Fritsch, M., Lynes, P. and Wood, D., eds., 2018. *Eco-deconstruction: Derrida and environmental philosophy.* New York: Fordham University Press.

Gaard, G., Estok, S.C. and Oppermann, S., 2013. *International perspectives in feminist ecocriticism: making a difference.* London: Taylor & Francis Group.

Ghosh, A., 2016. *The great derangement: climate change and the unthinkable.* Chicago, IL: University of Chicago Press.

Gibbs, L.M., 2006. Valuing water: variability and the Lake Eyre Basin, Central Australia. *Australian Geographer*, 37, 73–85.

Gomez-Barris, M., 2017. *The extractive zone: social ecologies and decolonial perspectives.* Durham, NC: Duke University Press.

Grosz, E., 2017. *The incorporeal: ontology, ethics and the limits of materialism.* New York: Columbia University Press.

Hamilton, C., 2017. *Defiant Earth: the fate of humans in the Anthropocene.* Sydney: Allen and Unwin.

Haraway, D., 2016. *Staying with the trouble: making kin in the Cthulucene.* Durham, NC: Duke University Press.

Heise, U.K., 2017. Introduction: planet, species, justice – and the stories we tell about them. *In*: U.K. Heise *et al.*, *The Routledge Companion to the Environmental Humanities.* London: Routledge, 1–10.

Hennessy, R., Cothren, A. and Matthews, A., 2022. Creating new climate stories: posthuman collaborative hope and optimism. *TEXT*, 26, 1–20.

Ingold, T., 2011. *Being alive.* London: Routledge.

IPCC, 2022. *IPCC assessment reports* [online]. Available from: www.ipcc.ch/about/ [accessed 7 November 2022].

Iovino, S. and Oppermann, S., 2014. Introduction: stories come to matter. *In:* S. Iovino and S. Oppermann, eds. *Material ecocriticism.* Bloomington, IN: Indiana University Press, 1–17.

Janke, T., 2019. *Protocols for using First Nations cultural and intellectual property in the arts.* Pyrmont, NSW: Australia Council for the Arts.

Jarvis, T., 2014. *Contesting hidden waters: conflict resolution for groundwater and aquifers.* New York: Routledge.

Johns-Putra, A., 2018. The rest is silence: postmodern and postcolonial possibilities in climate change fiction. *Studies in the Novel*, 50, 26–42.

Jolley, E., 1986. *The well.* Camberwell, Vic: Penguin.

Krauth, N., 2011. Evolution of the exegesis: the radical trajectory of the creative writing doctorate in Australia. *TEXT*, 15(1).

Ladson, A., 2008. *Hydrology: an Australian introduction.* South Melbourne: Oxford University Press.

Leap, D.I., 1999. *The handbook of groundwater engineering.* Boca Raton, FL: CRC Press.

Linton, J., 2010. *What is water? The history of a modern abstraction.* Vancouver: UBC Press.

Linton, J. and Budds, J., 2014. The hydrosocial cycle: defining and mobilizing a relational-dialectical approach to water. *Geoforum*, 57, 170–180.

Macfarlane, R., 2019. *Underland: a deep time journey.* London: Penguin Random House.

Macquarie Dictionary, 2010. *Macquarie dictionary*, ed. S. Butler. 5th ed. Sydney: Macquarie University.

Massumi, B., 2010. The future birth of affective fact: the political ontology of threat. *In:* M. Gregg and G. Seigworth, eds. *The affect theory reader.* Durham, NC: Duke University Press, 52–70.

McKinnon, C., 2023. Braiding. *In:* D. Wardle *et al.*, eds. *A to Z of creative writing methods.* London: Bloomsbury, 16–19.

Mertoja, H., 2017. *The ethics of storytelling: narrative hermeneutics, history and the possible.* Oxford: Oxford University Press.

Michael, M., *et al.*, 2011. *Social research to improve groundwater governance: literature review.* Report no. 66. Albury: Institute for Land, Water and Society, Charles Sturt University.

Mills, J., 2022. Dystopia creep. *Meanjin*, 80, 38–46.

Neimanis, A., 2017. *Bodies of water: posthuman feminist phenomenology.* London: Bloomsbury Academic.

Neimanis, A. and Walker, R.L., 2020. Weathering: climate change and the 'thick time' of transcorporeality. *Hypatia*, 29, 558–575.

Nelson, R., Barnett, S. and Kollmorgen, A., 2020. The evolution of groundwater management policy in the states of Australia. *In:* J.-D. Rinaudo *et al.*, ed. *Sustainable groundwater management.* Cham, Switzerland: Springer, 129–141.

Novicky, O., Kasparek, L. and Uhlik, J., 2010. Vulnerability of groundwater resources in different hydrogeological conditions to climate change. *In:* M. Taniguchi and I. Holman, eds. *Groundwater response to changing climate.* London: CRC Press, 1–10.

Pagnol, M., 1962. *The water of the hills.* London: Pan Books.

Paterson, B., 1896. Song of the artesian water. *Poetry Library* [online]. Available from: www.poetrylibrary.edu.au/poets/paterson-a-b-banjo/song-of-the-artesian-water-0004015 [accessed 16 August 2022]

Phillips, L. G., and Bunda T 2018. *Research through, with and as storying.* 1st ed. Abingdon: Routledge.

Plumwood, V., 1993. *Feminism and the mastery of nature.* London: Routledge.

Prose, F., 2012. *Reading like a writer: a guide for people who love books and for those who want to write them.* London: Aurum Press.

Rendell-Short, F., 2015. How the how: the question of form in writing scholarly works. *New Writing*, 12, 91–100.

Richard-Ferroudji, A. and Lassaube, G., 2020. The challenge of making groundwater visible: a review of communication approaches and tools in France. *In:* J.-D. Rinaudo *et al.*, eds. *Sustainable groundwater management: a comparative analysis of French and Australian policies and implications to other countries.* Cham, Switzerland: Springer, 191–209.

Rigby, K., 2009. Writing in the Anthropocene: idle chatter or ecoprophetic witness. *Australian Humanities Review*, 47, 1–10.

Ritson, K., 2019. *The shifting sands of the North Sea lowlands: literary and historical imaginaries.* Abingdon: Routledge.

Rockström, J., *et al.*, 2023. The what, why and how of the world water crisis: Global Commission on the economics of water Phase 1 review and findings. Global Commission on the Economics of Water. Available from: https://watercommission.org/wp-content/uploads/2023/03/Why-What-How-of-Water-Crisis-Web.pdf [accessed 18 March 2023].

Rose, D.B., 2013. Slowly – writing into the Anthropocene. *TEXT*, Special Issue 20.

Rose, D.B., 2014. Arts of flow: poetics of 'fit' in Aboriginal Australia. *Dialectical Anthropology*, 38, 431–445.

Schneider-Mayerson, M., *et al.*, 2020. Environmental literature as persuasion: an experimental test of the effects of reading climate fiction. *Environmental Communication*, advance online publication. https://doi.org/10.1080/17524032.2020.1814377

Segal, M., 2017. The missing climate change narrative. *South Atlantic Quarterly*, 116, 121–128.

Seigworth, G.J. and Gregg, M., 2010. An inventory of shimmers. In: M. Gregg and G.J. Seigworth, eds. *The affect theory reader.* Durham, NC: Duke University Press, 1–25.

Sheldon-Collins, D. and Heiss, A., 2014. 'Getting it right': Anita Heiss on Indigenous characters. *Wheeler Centre*, 4 November [online]. Available from: www.wheelercentre.com/notes/221927959a6b [accessed 14 February 2018].

Sparks, C. and Michael, R., 2018. Why speculate – the current state of 'spec-fic' publishing. *TEXT*, Special Issue 51, 1–15.

Stevens, L., Varney, D. and Tait, P., eds., 2018. *Feminist ecologies: changing environments in the Anthropocene.* Gewerbestrasse, Switzerland: Palgrave Macmillan.

Sussex, L., 2018. The end of the world as we know it. *Sydney Review of Books*, 17 September [online]. Available from: https://sydneyreviewofbooks.com/dyschronia-jennifer-mills-how-to-bee/ [accessed 22 September 2018].

Tredinnick, M., 2005. *The land's wild music: encounters with Barry Lopez, Peter Matthiessen, Terry Tempest Williams, and James Galvin.* San Antonio, TX: Trinity University Press.

Trexler, A., 2015. *Anthropocene fictions: the novel in a time of climate change.* Charlottesville, VA: University of Virginia Press.

Tynan, A., 2020. 'Que peut la fiction?' Storying the unexperienced experience in Jorge Semprun's fiction. *Modern Language Review*, 115, 46–62.

Vannini, P., 2015. Non-representational ethnography: new ways of animating lifeworlds. *Cultural Geographies*, 22(2), 317–327.

Wardle, D., 2019. *Storying with groundwater: why we cry.* PhD thesis. RMIT University.

Wright, A., 2006. *Carpentaria.* Sydney: Giramondo.

Wright, A., 2016. What happens when you tell somebody else's story? A history of Aboriginal disempowerment. *Meanjin*, 75, 58–76.

Wright, A., 2022. The inward migration in apocalyptic times. *Meanjin*, 81, 57–65.

2

THE WATER TABLE

WAYS OF KNOWING AND NOT KNOWING GROUNDWATER IN THE ANTHROPOCENE

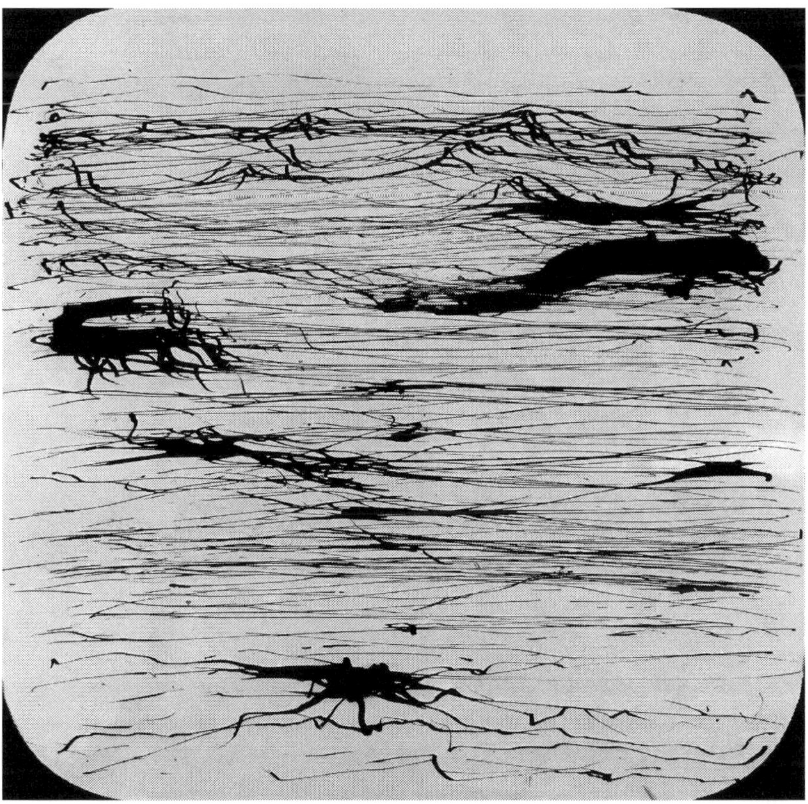

Figure 2.1 Bill Sampson, *Lines ground poetry*, 2020, ink and enamel on aluminium, 20 × 20 cm.

DOI: 10.4324/9781003270416-2

The water table is the subsurface boundary between unsaturated (or vadose) subterranean material (rocks, gravels and sands) and the deeper saturated (or phreatic) materials. What is called a 'capillary fringe' separates these zones, the changeable boundaries of the water table. A more technical definition of the water table specifies that it is the subterranean water surface where the water pressure in the rock/soil pores equals atmospheric pressure. Above the water table air and water occupy the rock/soil pores, and below the water table the rock/soil pores have no space for air. Rock formations saturated with water below the water table are called aquifers.

> The fluid – as that other, inside/outside of philosophical discourse – is inherently unstable. Unless one subjects it to geometrisation or (?) idealizes it.
>
> Irigaray 1985, quoted in Volkart 2022, p. 1

In this chapter I consider ontological and epistemological perspectives, from which and into which groundwater narratives flow. In the quote above, Irigaray points to the mutable relations between humans and groundwater, that which can be known and not-known, the realm of subterranean imaginaries. The water table fluctuates as various forces of aquifer recharge and discharge interact. Recharge means the seeping of rainwater into aquifers or water entering aquifers from the beds of lakes, creeks and rivers. A plethora of discharge activities removes groundwater from subterranean domains. These include water leaving aquifers through natural springs, through movement of groundwater into swamps, lakes, creeks and rivers, and increasingly the various human-made structures and devices for removing and pumping water from aquifers.

The movements of groundwater are called flux. Flux in hydrogeology is a unit of measure revealed by Darcy's law – the flow of water in a unit of area, measured in distance over time. Flux, sometimes known as hydraulic conductivity, is more than the rate of flow, because it takes account of the volume of the area through which the flow occurs, and the effects of time, gravity, and hydraulic head or pressure on the flow (Fitts 2013, pp. 49–52). I am guided towards expressing partialities of groundwater's materiality, as what I know and do not know move along the slippery slide of understanding what constitutes 'water' in a material, globally warming world. The range of influences on groundwater flows, or the ways that groundwater is in constant three-dimensional movement in relation to its surroundings, flux, can be seen as an analogy for writing processes. Discursive relations to water give me a place to write from. As Jamie Linton explains:

> The very process that we call 'life' is so conditioned by water that it becomes impossible to separate these two processes except by fixing them in abstractions such as 'people' and 'H_2O.' Thus, knowing and identifying water is necessarily a product of engagement, with engagement itself being the real – that is, relational – substance of both the knower and the known.
>
> Linton 2010, p. 223

In the context of this book, climate writers engage with narrative groundwater to come to know its relational expressions. We know groundwater as a product of our engagement with its many forms and expressions.

Giving attention to what we can know, and how we might know it through literature, has been a long endeavour. Mid-twentieth century modernist thinking questioned writers' capacities to represent external realities mainly because of the subjectivity of individual perspectives. Modernist views were challenged by postmodernist approaches, which acknowledged the uncertainties of what language can convey (Boulter 2007, p. 77). Neither objective nor subjective realities were thought to be fixable in text. Building momentum in the twenty-first century, posthuman and new materialist theorising turned to considering the relationality of matter, aiming to unsettle the anthropocentrism of earlier approaches. It is this turn away from human-centredness and towards an acknowledgement of matter's vitality that supports how we might come to know narrative groundwater. The frameworks of thinking and the concepts writers adopt influence the ways that we might construct groundwater narratives. This chapter discusses how our thinking about the Anthropocene affects how we might imagine groundwater through climate writing. Climate writing, I argue, is melded to how climate writers come to know of and think *with* the materiality of their topic.

Knowing and Not Knowing

Cognition of underlying ontologies and epistemologies underlies creative work. Articulating ways of thinking about the beingness of 'subjects' – whether the 'subject' includes elements of environmental magnitude or smallness, environmental degradation or extinctions, or the specific aspects of climate catastrophe being narrated – gives depth and context to the writing process. While writers need not be scholarly philosophers, understanding the epistemologies and ontologies of the 'things' they write about/with is essential. Climate writing means engaging with nonhuman entities, establishing a relationship with the matter we are writing about/*with*. A wide range of 'matters' are possible – lightning, food, metals, the air we breathe, floods, droughts, endangered and extinct species or any particular effects of global warming.

Articulating the ways of knowing the material qualities and agencies of climate writers' subjects also implies, in its reverse, explicating the state of not knowing throughout various stages of the writing. Self-consciously expressing ways of knowing and not knowing draws attention to how climate writers approach what might start as or become uncertain or doubtful through the writing. It is difficult to comprehend groundwater as a fixed entity, because it is virtually impossible for groundwater to be delimited in space and time. Hydrogeologists' descriptions are always provisional; groundwater's transience is its norm (Currell *et al.* 2016). Groundwater narratives, likewise, open avenues to exploring what may not yet be known.

Not knowing, combined with curiosity, is often the opening gambit to a writer's research and writing practice. As Julienne van Loon (2023) describes, the common feeling of not knowing the core or the trajectory of one's writing, particularly in the early or middle stages of a large literary work, is an integral part of imaginary processes. Deborah Bird Rose cites her favourite philosopher, Lev Shestov, who argues 'against the great "crime of humanity", which is to banish our capacity to encounter the unknown without subjecting it to human justification' (Rose 2013, p. 11). The living world, Rose goes on to argue, 'is not knowable as a whole by any of its constituent parts' (2013, p. 11). I start with not knowing, trying not to justify my uncertainty, but sitting with the discomfort. Many aspects of groundwater flows are at present unknowable. The potency of not knowing, Rose argues, is that 'acknowledgment of ultimate unknowability draws attention to how we punctuate and frame our questions. … Fidelity to that which cannot be fully known is absolutely integral to slow ethics' (2013, pp. 11–12). Through experimentation and unfolding iterations of imaginations of groundwater's real and metaphoric potency, I immerse myself in uncertainty and doubt, playing with and unsettling the privilege of certain knowledges. I write into the fog of my amateurish knowledge of groundwater, uncomfortable with the risk of errors, listening for the voices and agential qualities of groundwater. Not knowing remains one of the writer's tools in the grapple with how we come to partially understand the ecological materials we write with. It is not only not knowing the labyrinths of hydrogeology; my thinking *with* groundwater traverses not knowing how narrative groundwater will speak up for itself.

Knowing and not knowing groundwater are influenced by it being predominantly an invisible subject. Swiss abstract artist Paul Klee argues that the role of painting is to render the non-visible, rather than representing the visible (Daley 2014, p. 618). Luce Irigaray, the water- and fluid-focused feminist philosopher, illuminates the invisibility of the 'maternal-feminine' and refers to 'invisible' as 'relations between us and the world, us and the other(s)' (Irigaray 1993, quoted in Daley 2014, p. 618). Rendering expressions of invisibility in subterranean imaginaries aims to expand a discourse around groundwater's roles in the hydro-commons. I contend that storying *with* so-called 'invisible' groundwater invites an examination of the relationality between humans and groundwater. In the Anthropocene this relationality is ripe for exposure.

While the water table presents a solid skin-like image, a surface likened to a tabletop, a plane that represents a division between two levels of saturation of rock with water, it is probably none of these. The capillary fringe is the hydrological term to indicate the aquifer's movements at the level of the water table. I imagine the water table more like a floating scarf, a ceiling-like fabric in a Bedouin tent, where below the air is thick with aromatic shisha pipe smoke and above the scarf the air is lighter, breathable. The wavering ceiling fabric might be imagined as a fast-motion action of the water table as it responds over days, weeks, months or centuries to water inflows from rain (called recharge of an aquifer), and to water outflows from bore pumping or flows into rivers, lakes or the ocean (called discharge from an aquifer).

Karan Barad uses the metaphor of lightning to open her trans-material exploration of 'the nature of matter and its agential capacities for imaginative, desiring, and affectively charged forms of bodily engagements' (2015, p. 388). Barad reminds us of Mary Shelley's galvanising interest in lightning as a source of life for her protagonist Frankenstein. Writers thinking about/*with* groundwater explore its metaphorical potency, its means of flowing through a story, from epistemological foundations. Groundwater may not be as dramatic as a spark of lightning, but its slow-moving subterranean plumes, following crevices, filling microscopic pores, carrying uniquely endemic stygofauna through life cycles, transporting minerals perhaps across thousands of kilometres over centuries, are all part of groundwater's agential capacities. Groundwater has a plethora of characteristics that feed its narrating agencies.

Like Linton and Budds (2014, p. 173), I start with understanding water and its relation with society through a consideration of dialectical relations, not so much as *between* humans and water, but as the relations that *constitute* both water and human societies through dynamics of social power.

Applying this idea to many instances of groundwater's relationality with human societies, it is clear to see that transformations in the hydrogeological environment (say the engineering capacities to pump deeper bores, to access older aquifers for agriculture and irrigation) produce changes in society (say the development of communities, cultural practices and meanings of groundwater in places that have little surface water). In turn, these iterations bring changes in many aspects of how water is understood and 'managed'. This iterative cycling of internally related features of groundwater and societies, where both continue to make and remake each other, reveal many and varied 'geometries of power' (Linton and Budds 2014, p. 173). Groundwater plays an increasingly pivotal role in the hydrosocial cycle – as its value in social and cultural systems becomes evident. As mentioned in the first chapter, referring to the water cycle, when groundwater emerges at the surface it is used in various ways by humans and nonhuman ecosystems. What climate writers know and do not know of groundwater in any of these states of becoming influence the processes of storying *with* groundwater.

New Materialist Thinking and Becoming

As many new materialist thinkers have argued, matter matters in addressing the reality of global warming (Bennett 2010; Haraway 2016; Barad 2017). New materialist thinkers and writers pay attention to human entanglements with the nonhuman world, leading to posthumanist interests in ethics and politics (Jagodzinski 2021). New materialist thinking provides a way into thinking with matter, engaging with the vitality and agency of entities that have previously been considered to be passively outside the human realm. Groundwater is one such entity that has been considered in the Western world, and historically across many cultures, as a resource to fuel economic growth and development of human capital. Creative writing based on new materialist entanglements between humans and the waters that keep us alive tests this viewpoint. Common complacency about existent and looming climate catastrophes is also challenged.

Many cultural theorists are expanding understandings of human relationships to matter and more-than-human entities.[1] Articulating the intra-actions between discourse and matter is all the more relevant in the Anthropocene because of the urgency of stories that need to be told.

Material phenomena are seen as 'knots in a vast network of agencies, which can be "read" and interpreted as forming narratives, stories' (Iovino and Oppermann 2014, p. 1). As Iovino and Opperman go on to say:

> Developing in bodily forms and in discursive formulations, and arising in coevolutionary landscapes of natures and signs, the stories of matter are everywhere: in the air we breathe, the food we eat, in the things and beings of this world, within and beyond the human realm. All matter, in other words, is a 'storied matter.' It is a material 'mesh' of meanings, properties, and processes, in which human and nonhuman players are interlocked in networks that produce undeniable signifying forces.
>
> Iovino and Oppermann 2014, pp. 1–2

Rhizomatic becoming is another way to 'know' groundwater in its various internal states (Deleuze and Guattari 1988, pp. 300–305). Creating new subjectivities through networked and collective actions of language, or rhizomes of meaning, is how Deleuze and Guattari (1988) proposed posthuman ideas of otherness and difference. Connected networks of meaning, like subterranean rhizomes, enable new subjectivities and differences to become newly affirmed. Groundwater is becoming itself in an aquifer or becoming another part of the water cycle as it moves, either infinitesimally slowly through sandstones, or with force through large crevices and caves. It is not only the language of hydrology and hydrogeology that populates groundwater narratives. Chemistry shows me that water is not simply H_2O in stasis. Water is in constant flux. In any body of water H_2 molecules and separate hydrogen and oxygen atoms float, merging, becoming H_2O, splitting to form H_2 and O_2 molecules, and re-becoming H_2O again. H_3 will occasionally, momentarily form if water has a strong electric current passed through it. Water, when considered as lively matter, is constantly in the state of becoming. In storying *with* narrative groundwater, groundwater becomes the fluid interstitial space, the go-between. Groundwater becomes the 'between': between itself and me, between the writer and the reader. Groundwater is literally the connective fluid between human and nonhuman ecologies and geological spaces and matter. Water, as atomic matter, two hydrogen atoms bound to and unbound from oxygen atoms, plus all the salts and minerals it captures and contains, moves and deposits, is becoming. Narrative groundwater as connective fluid is becoming. When it is written in narrative form, likewise it is always between and becoming. Groundwater reserves are rarely static; rather they are constantly somewhere between recharge and discharge, always becoming something different as they come under the forces of gravity and move to lower places, influenced by the forces of geological pressures and the chemistry of the rocks through which they move. Narrative groundwater is a place of flux – taking several meanings of the word. It is liquidity, fluidity. It is change and permutation. It is flow and movement.[2]

Humans may come to understand the ongoing becomings of worldly matter through the intra-actions of material and discursive practices. Donna Haraway has called the networks of real bodies of matter and the discourses that are generated with/about them our 'material-semiotic reality' (2004, p. 2). In this light, and influenced by Deleuze and Guattari's influential text *A thousand plateaus* (1988), thinking of groundwater's beingness in the ontological mode of 'becoming' enables a new vocabulary within groundwater narratives. A focus on the transitory differentials between groundwater's states of being, rather than expecting solidity or a static state of being, shifts the bedrock to subterranean imaginaries.

Building on the idea that matter is always in a state of becoming, Karen Barad (2015) explains how electrons behave in matter, as opened up by quantum physics and quantum field theory. She draws comparisons between new ways of understanding matter and less determinate ways that artists and writers can know and create meaning in their work. As Barad states: 'Ontological indeterminacy, a radical openness, an infinity of possibilities, is at the core of mattering' (2015, p. 401). Human and nonhuman matter is perceived as being materialised through intra-action of the entities and discourse, and subject to the same ways of becoming and being. This breaks the previously dominant pattern that holds humans as souled and sentient, separate from and able to control and dominate the supposedly predetermined, nonhuman world (Coole and Frost 2010, p. 8). With this in mind, creating an imaginative discourse of groundwater's expressions is an acknowledgment of its vitality. Expressions of groundwater in creative writing are imagined through its indetermined beingness, its wateriness, on a continuum with not only human characters and nonhuman ecologies, but also in intra-action with the author, the writer. The notion that 'things' are not fixed entities, but always in the process of 'becoming' holds great potency for climate writers' narrative expressions of the impacts of living in the Anthropocene.

Material ecocriticism has built on the 'material turn' to analyse literature through the emerging non-anthropocentric paradigm. It focuses on 'networks of bodies and discourses' (Iovino and Oppermann 2014, p. 2). Drawing on ideas from material ecocriticism opens the possibility of examining how climate fiction writing might interact with nonhuman entities – in my case, the material nature and forces of groundwater – 'both *in* texts and *as* text' (Iovino and Oppermann 2014, p. 6). Coole and Frost (2010, p. 3) argue that we have perhaps lost sight of matter itself, and its capacities, not just of being, but of acting in a material world. At the same time, ever-emerging scientific understandings, from quantum physics to biological and ecological knowledge, open questions about the nature of matter and human beings' relations to the slippery slide of what constitutes a material world (Barad 2007). If 'forces, energies and intensities' are the language of materialist viewpoints (Coole and Frost 2010, p. 13), what are the effects on the expressions I give to groundwater and to the human characters who experience it in their lives? 'Reading' and writing groundwater as agential matter – in contrast to describing subterranean aquifers as representable places – enables me to establish groundwater's resilience as well as its fragility. Finding ways to voice or express the material reality and literary potency of extra-large, moving bodies of groundwater means thinking into the actions of groundwater. It is the self-organising properties of narrative groundwater that seek expressions in creative forms.

Guattari's 'ethico-political articulations – *ecosophy* – between the three ecological registers (the environment, social relations and human subjectivity)' (2000, p. 28) provide an effective vocabulary for dealing with threats and transformations on planet Earth. The political and agential qualities of matter emerge not only through giving attention and subjectivity to matter, but through caring for things, building a capacity to feel care for the natural world (Herbrechter and Callus 2008). Giving attention to groundwater, and caring for its sustainability and capacity to maintain ecologies, includes building ethico-political thinking into climate writing practices. Narrative groundwater asks climate writers to pay attention as they facilitate stronger, more ethical, more inclusive and more politically affective relationships between humans and nonhuman ecologies.

Rather than perceiving groundwater as discrete, inert and inherently measurable, a new materialist approach applies a non-static ontology that perceives 'excess, force, vitality, relationality, or difference that renders matter active, self-creative, productive, unpredictable' (Coole and Frost 2010, p. 9). The narrative potential in this view of matter means challenging the view that stories come from humans. Rather, groundwater narratives emanate from the interrelated agencies of both the humans and the nonhuman entities that populate the narrative. The capacity for groundwater to 'speak', either through the characters' observations of the effects of its actions, or by expressing its own volition, emerges from an ontological understanding of groundwater as mattered and agential. Highlighting the ways storied matter can act (Iovino and Oppermann 2014, pp. 2–3) means considering how things, entities or matter like groundwater exercise agency. Paradoxically, humans know of or intuit such agency through human interventions, including observations, imaginations, experiences and creating material texts. As an inconclusive exemplar of these ideas, readers of *Why we cry* learn of groundwater's corrosion of irrigation pipes, its movements in deep leads, its withdrawal from ecosystems. The mysteriously lowering water table is responsible for the death of the protagonist's horse, an inciting event early in the manuscript. Groundwater's intra-relationality seeps steadily through the narrative, increasingly showing its affective force.

According to Barad (2008, p. 140), an account of the materialisation of all bodies, human and nonhuman, active things, entities and objects, draws upon the integration of linguistic or discursive acts and material reality; mutually implicated meanings and matter's reality intra-act and result in the active agency of ongoing materialisation. Barad's notion of 'agential realism' challenges representationalism. She focuses on the intra-action of material and discursive practices through an ontology of posthuman performativity (Barad 2008, p. 129). Barad's discussion of performativity is a keystone. She sees performativity as a 'contestation of the unexamined habits of mind that grant language and other forms of representation more power in determining our ontologies than they deserve' (Barad 2008, p. 139). Following Judith Butler and Donna Haraway's lead, Barad (2008, p. 139) links performativity to the formation of the subject and to the production of the matter of bodies. Rather than representation, which 'separates the world into ontologically disjointed domains of words and things' (Barad 2008, p. 140), Barad's posthuman performativity emphasises iterative intra-actions between the boundary-making function of words and matter. Here I consider groundwater narratives as a discursive practice. How climate writers know of their intra-actions with groundwater becomes the making ground of 'agential realism' (Barad 2008, p. 129). Creating a realistic view of groundwater's agential potential ripples beneath subterranean narratives. Delving into intra-relational ontologies is a significant way to know and not know. It opens the door to subterranean imaginaries that perform the agential realism of groundwater.

Material Eco-feminism

A material feminist lens on writing practice is another important way of knowing and questioning groundwater's material impact. Acknowledging the intra-actions and agency of a material, watery world, rather than seeing it only as a construct of language, is a significant aspiration of an eco-feminist standpoint (Braidotti 2013; Haraway 2016; Neimanis 2017). Leaning heavily on the strong tradition of Australian eco-feminist thinking, several key feminist principles emerge to assist reconceptualising the material aspects of groundwater. These include: linking global issues with local ecologies; expounding the personal as political; and shifting from a gender-blind posthumanism to an acknowledgement of material, embodied and gendered elements of human and nonhuman enmeshments (Rigby 2018; Stevens, Varney and Tait 2018; Rose 2014; Plumwood 2018). Feminists have long argued that humans, nonhumans, technology and natural agents all jointly interact to mutually construct meaning in matter (Alaimo and Hekman 2008a, pp. 5–6; Barad 2008; Haraway 2016). Within binary approaches to nature/culture, women are often associated with the natural world, another 'thing' for patriarchy to control, and over which to exercise mastery. Second-wave feminists argue strongly that the material effects of discriminating, misogynist elements of patriarchal culture must be taken seriously.

Eco-feminism redresses the limitations of a singular focus on the discursive over the material, avoiding the separation of language, discourse and culture from materiality. Critiques of positivist perceptions of a so-called objective reality also have roots in feminist analysis (Alaimo and Hekman 2008a, pp. 3–5). Analysing gendered power relations, as practised through language, discourse and views of materiality, influences how writers give voice to human and more-than-human characters. Lifting a feminist and queer lens to knowing and not knowing groundwater's many expressions disrupts the assumption that language can represent what is real in a gendered world.

An eco-feminist approach to writing with groundwater recognises the risk of ascribing female qualities to the natural world (Stevens, Tait and Varney 2018). The 'land-as-woman' metaphor can be seen to support the normalising of ongoing exploitation and aggression towards both women and environments (Kolodny cited in Glotfelty 1996, p. xxix). Groundwater narratives are wary of metaphors of 'mother earth', 'mother nature' or 'mother water' to express reductive views of femaleness. Neimanis's concept of gestational water provides a feminist take on water ontologies, insisting that feminine seas and amniotic waters are not '*like* planetary waters, but continuous with them' (Neimanis 2017, p. 113). Through her concept of gestational water, Neimanis avoids essentialising women's experience, nor does she assume that women belong more naturally to 'nature'.

Material feminism opens potential for an ethics based on the material consequences of human and nonhuman interactions. Applying a gendered analysis of patriarchal influences on how matter comes to matter quickly shows how embodied, situated actions are based on unequal and often unjust power relations. Political consequences of gendered analyses are seen as scripted onto human and nonhuman bodies (Alaimo and Hekman 2008b). Under the broad eco-feminist umbrella, material feminism focuses on a feminist environmental politics that takes account of environmental sciences, and places this knowledge about the so-called 'natural' world, not as mere background, but as a material actor in the realms of science and feminist politics (Rigby 2018; Stevens, Varney and Tait 2018). By beginning with the voices of matter, unexpected social and political alliances are possible. Expanding a materialist eco-feminist ethics and politics to climate writing is a firm foundation to groundwater narratives that address climate change with a political agenda for social and cultural change.

Challenging Anthropocentrism

Climate writing influenced by the posthuman turn aims to de-centre the human subject and focus on the material agency of things. Conversations and collaborations between sciences and ecocritical humanities disciplines increasingly analyse non-anthropocentric approaches to language and reality, and contribute to matter finding its own voice (Iovino and Oppermann

2014, p. 2). Although the Cartesian derived nature–culture dualism does not have a sound ontological foundation, this separation has become a 'known' reality simply through historic usage (Álvarez-Miranda 2021, p. 138). Ironically the traditional imagined dualisms are starting to dissolve because of the recognition of the damage caused by a fractured perspective that normalises human dominance of nature (Álvarez-Miranda 2021, p. 138). Also ironically, at the same time as Western and capitalist culture's responsibility for climate catastrophes is being slated home, scientists from advanced economies are progressing enormous scientific and technical innovations that give us to some extent the means of addressing many of the planet's ecological problems and social inequities. Paradoxically though, humans appear to lack the social or cultural forces to enact adequate solutions (Álvarez-Miranda 2021, pp. 137–138).

The belief that anthropomorphising 'nature' (describing 'natural' beings and entities with the characteristics usually ascribed to humans) is problematic is being questioned. Many now argue that ascribing human characteristics to animals, mountains, rivers and things opens the potential for humans to understand, care for and protect 'nature' (Plumwood 2008, 2009; Bennett 2010, p. 98; Le 2017). However, anthropomorphic metaphors may risk placing groundwater as the backdrop to human endeavours, or may tie the writing to gender-blind, anthropocentric positioning. Conceptions of power and emancipatory ideals must not be lost as the 'sins' of anthropomorphising are challenged. Narrative groundwater chooses its metaphorical referents carefully. Evoking the metaphorical parallels between groundwater and human fluids, the idea of interiority gestures towards the value of caring for self and environmental waters.

Escaping the pulls of human-centredness in posthuman styled climate writing presents conundrums. Humans are 'disappearing' (no longer the central actors) at the same time as human agency is most needed to act with response-ability for the damage we have created. Disruption to a human-centred narrative form by giving expression to ecological forces, however, paradoxically includes acknowledging and challenging the writer's hegemonic authorial position. Disrupting the idea that nonhuman material entities can and should be silenced remains important. Climate change writers blur the lines between beginnings and endings of the human lives and the ecologies they share. Climate change narratives aim to 'model desirable ecological relationships and values indispensable to environmental change'

(Kostkowska 2013, p. 4). They might also demonstrate the consequences of human-centred standpoints, where environmental impacts are ignored.

Climate writing looks for resonances between the 'many conative actants swarming and competing with each other' for a voice (Bennett 2010, p. 122). It gives expression to unruly actants and agential entities and beings from the nonhuman realm. Human and nonhuman characters create a polyphony of perspectives. Polyphonic voices of both humans and nonhuman entities provide another framework for the construction of groundwater narratives. There is no singly correct 'voice' to tell a story; it is from the social heteroglossia of possibilities that groundwater narratives emerge (Bakhtin 1981; Boulter 2007, p. 67). Human writers may start with the notion that we can know the world through its expressions (Massumi 2001, p. xvii). Climate writers become the conduit for expressions of a messy, polyvocal material world.

A conundrum of writing *with* rather than about 'nature' and 'things' from an enmeshed emplacement in the world lies in the humanness of the writing process. Writing from the entanglements of anthropocentric language, into posthuman perspectives, entails knowing, feeling and imagining the writer's enmeshments with their subject. Freya Mathews's concept of 'ecological self' gives a clue to how to think *with* the entities with which we are enmeshed. The 'ecological self':

> calls for alternative epistemologies that pull us into communicative and Lawful engagement with reality rather than distancing us from it as strictly disinterested observers. This is by no means of course to suggest a wholesale repudiation of science – nothing could be less desirable at a time, such as now, of raging populist anti-rationalism. But it does invite us to open up the 'imaginary', so to speak, of reason, so that conventional rationalist disciplines, such as science, may be integrated with alternative cognitive modalities.
>
> Mathews 2021, p. xxviii

I discuss integration between groundwater narratives and science later in this chapter and in more detail in Chapter 4.

It is not just that it is water – it is the qualities groundwater assumes by being multi-aged, underground and ever-cycling, becoming and unbecoming, that fuel its narrative expressions. Climate writers listen to the voices of water as a guide to what can be known and not known. In *Thinking with water* the editors/authors seek to cultivate progressive ways of relating to water that apply well to thinking *with* groundwater:

> Water teaches us about an ethics of care and response; about the need for all bodies to be sustained by good water, good nourishments, good words; about patience and resilience; about the way theory, activism and art overlap and blur; about the need to listen to voices that are too often silenced or deemed mere 'babble'.
>
> Chen *et al.* 2013, p. 19

Climate writers listen towards what would otherwise not be considered. Groundwater's voices, subterranean imaginaries, remain partial, obscured. In groundwater narratives, the writer's task is more to illustrate the elusiveness of groundwater through writing its unknowability, its relationality and affects, than to doggedly portray its characteristics (Vannini 2015, p. 16). Anthropocentrism is unsettled in the process.

Narrative Groundwater as a Hyperobject

Timothy Morton's (2013a, 2013b) concept of the *hyperobject* gives me a way to scaffold groundwater's voices in climate writing. Hyperobjects provide a way to think about narrative groundwater's hidden scale, its relationships and movements. Morton's hyperobjects address worldwide phenomena and 'massively distributed entities that can be thought and computed, but not seen or touched' (Morton 2013b, p. 37). We might consider a spreading virus, like the COVID-19 pandemic, something of a hyperobject. Morton argues that, by thinking outside the human, 'every entity renders the real in some sense according to its physical form and structure' (2013a, p. 8). Raindrops to oceans, mice to mountains, bacteria to viruses, all render their reality with their own agency within the hyperobject of global warming, according to Morton.

Applying the idea of hyperobject as a way of thinking *with* groundwater behoves thinking of aquifers beyond being a singular object. Groundwater, as a hyperobject, occupies multiple spaces, multiple relationships, with movement over ancient timeframes. Imagining the viscosity and non-locality of groundwater as a hyperobject becomes a partially helpful way of building groundwater narratives. Morton bases the concept of hyperobject on a philosophy that rejects the primacy of human reason over experience.

Thinking about hyperobjects starts with an acceptance that any separation between humans and nonhumans has been truly dismantled. We are all in the soup together, from minuscule to enormous.

Narrative groundwater's presence can be partially rendered by teasing out the five interrelated qualities Morton ascribes to hyperobject: viscosity, non-locality, temporal undulation, phasing, and finally their interobjectivity.[3] First, considering the viscosity of hyperobjects means considering the stickiness of things, that objects are there in their vital beingness (Bennett 2010). Second, thinking of the non-locality of groundwater also rings true with the disparate spread of aquifers throughout regions and landscapes: 'Hyperobjects cannot be localized' (Morton 2013b, p. 40). Morton uses the analogy of swimming in a pool, the 'everywhereness' of the water in the pool in relation to our independent bodies, and how each affects the other. Our bodies produce ripples and diffraction patterns of light in the water, and the water causes our skin to cool, to goose-bump and wrinkle if we stay in long enough (Morton 2013b, pp. 41–42). Similarly, I cannot think of groundwater as one place. It is in many places, an entity without a locale. When hydrogeologists map groundwaters' localities, and draw arrows to represent its movements, it is tempting to forget that these are merely blurry estimations of place. As Frankie says to her hydrogeologist friend Helen, in the following excerpt, 'Subterranean worlds can hardly be expressed in lines.' Like hyperobjects, aquifers 'are shared by numerous entities in a common sensual place' (Morton 2013b, pp. 41–42). I think of people soaking in hot springs, perhaps alongside extraordinary biota. Groundwater might be known as a shared space, 'a vast non-local configuration space' that Morton calls 'the mesh' (2013b, pp. 40–41). These viscous, non-local qualities of groundwater seeped into my storying.

Third, groundwater can be perceived as multi-aged. Imagine the first minutes water seeps into semi-saturated subsoils, and at the same time moves through aquifers' deeper realms. The same aquifer may be thousands of years old, having flowed slower than imaginable through sponged crevices. This sits with Morton's notions of temporal undulation, or molten temporality: 'Any massive object distorts time-space' (Morton 2013b, p. 40).

Fourth, Morton also describes hyperobjects as 'phased'; they seem to come and go, like the seasons, and are partly invisible to humans. They are part of larger structures, multidimensional systems. Groundwater heaves under the influence of the moon, flows with the forces of gravity, and is squeezed by geologic pressures through incomprehensible timeframes. It is at first glance not visible. Yet, Morton (2013a) explains, hyperobjects continue to unfold elsewhere, whether we look or not. Groundwater may emerge with its own volition, its own pressure system, as a spring in one place, and elsewhere the same aquifer will be low and subtle, reluctant to be drawn to the surface, other than with the forces of deep centrifugal pumps. The coming and the going mean we humans are unable to see or hear the phasing, the full extent of groundwater's multidimensionality. We may see small 'indexes' of the whole. The index, like the list in the back of a book, indicates the parts of the hyperobject we can focus on. A flock of birds on a wetland swamp may not be just that, but an indexical sign of global warming, an index, cross-referenced to the receding subterranean water below the ducks, and to the water that emerges at another time and place.

Interobjectivity is Morton's (2013a, 2013b) fifth quality of *hyperobjects*. He describes interobjective reality as a sum total of all the 'footsteps' of many objects stepping into each other over time; it is therefore necessarily non-local. A dinosaur's footprint in mud becomes a hole in a rock seen by humans sixty-five million years later, creating a sensuous connection between the dinosaur, the rock and the human, despite their different time scales (Morton 2013b, p. 41). Understanding hyperobjects as both things and relations, in the process of phasing, where we may see only glimpses of their relational effects, assists thinking about how language might perform the presence and actions of groundwater in climate writing (Morton 2013a, p. 73).

Within the broad spectrum of a posthuman, new materialist or what Morton calls an object-oriented ontology, as distinct from a human-centred way of considering groundwater, I cannot write *about* a hyperobject. I can only attempt to 'evoke' hyperobjectivity through groundwater's multiple presences. Irony becomes more important than ever, argues Morton, as we humans find that we are not separated from a disappearing 'world' or 'environment'. Rather, ironically, we are glued, enmeshed more than ever to the hyperobjects we find ourselves within (Morton 2013b, p. 48). When writing *with* a hyperobject, from which I am not separated, efforts to express the more-than-human are partial and limited (Buell 2001). Rather than a coherent depiction of a singular place, incomplete depictions of narrative groundwater become imaginaries of groundwater's movements.

A risk in writing groundwater as a hyperobject is to inadvertently continue to reinstate groundwater as an external 'object', as if it is a finite 'thing'. The writerly perils of objectification of groundwater are apparent, particularly when immersed in a quantitative field of science such as hydrogeology. The challenge remains to think into a way of imaginative creativity that acknowledges, as did Deleuze, that the presence of even invisible matter is sensible, or able to be sensed, through its affects. Groundwater may not be physically visible, but through creative imagination I contend that its presence and affects can be expressed. Knowing and not knowing the various elements of narrative groundwater's hyperobjectivity builds its storying capacities.

Science Knowledge and Uncertainty

Science has often been the topic of literature, and writers increasingly turn to science for knowledge (Sleigh 2011; Segal 2017). At the same time, the stories we create contribute to knowledge (Sleigh 2011, pp. 4–5). It is widely argued that climate scientists need the arts, as a way to widen perceptions and knowledges of the effects and the affects of climate change (Thomashow 2002; Clarke and Rossini 2011; Sleigh 2011; Baucom and Omelsky 2017). The productive relationship between knowing hydrogeology and climate writing is particularly important in the current context of escalating effects of global warming. Climate writing explores how the two tasks of embedding facts and expressing an author's witnessing of features of the world into narrative are compatible endeavours (Sleigh 2011, p. 18). Groundwater narratives interrogate the zone between literature and science with the aim of showing the important relationship between these forms of knowledge. Climate writers rely on the knowledge of hydrogeologists to re-present calculated and estimated features of groundwater for readers. At the same time human characters may be sceptical or outright critical of science's claims to overarching knowledge (Buell *et al.* 2011, p. 422). They may also criticise the role of science and engineering in the creation of groundwater's vulnerable plight. Climate writers may question claims of certainty in hydrogeology and the science of climate change (Buell 2001; Baucom and Omelsky 2017). However, in the same breath, climate writing needs science to make credible descriptions and world building in convincing ways.

All the hydrogeologists I spoke with in the research for this book at some stage brought out a scrap of paper and a pencil or pen and drew a sketch of a landscape in profile to illustrate a feature of groundwater's movements. In the course of research and monitoring groundwater flows, hydrogeologists use mathematical formulae to test and develop models of groundwater's volumes, chemistry, locations and flows. Line drawings are ubiquitous in hydrogeology texts and websites. Numerous cross-sections of landscapes appear like cut layer cakes, representing the movement of water from the surface, through unsaturated zones to zones of saturation in sub-surface aquifers. How we 'represent' things has an influence upon what we might know about them. The limits of lines, the limits of texts and mathematical models, constrain the imaginable complexity, the blurriness and movements of groundwater through various mediums. Lines are a blunt starting point.[4]

Hydrogeological sketches and representations of groundwater as finite bodies of water derive from discourses of the Enlightenment and a Cartesian, binary way of thinking, which assumes that through empirical science humans are separate from and can know an objective outside world. Such representations imply that groundwater is a 'thing' comprised of confined and manageable zones, as if its capacity to respond to geological and meteorological forces, to sustain stygofauna, and to carry nutrients and pollutants at great depths, can be sketched with lines. The influences of gravity and of geological pressures on groundwater, such as folding, faulting and sedimentation, are minutely, perhaps even intimately, contained in sketches, mathematical formulae or complex models. In comparison, the uncertainties and partiality of what climate writers may express are emphasised and acknowledged when writers aim to intimate the 'becoming' nature of groundwater. Ecomimesis, or attempts to write mirror descriptions that evoke a writer's experience of the environment, like pencil sketches, are notably difficult (Morton 2007). Climate writing or environmental art that aims to be factual and imaginative at the same time raises dilemmas for how climate writers might portray features of climate change (Morton 2007, p. 193). I propose that bridging between knowledge and imaginaries is a key strategy in perceiving the voices of narrative groundwater and storying them into credible groundwater narratives. On the 'bridge' climate writers face the risks of slippage, being drawn into dualisms that divide humans irrefutably from groundwater's potency, its agency, its liveliness. The task is to traverse the slippery terrain of narrative groundwater as a becoming subject. We may hold our footing momentarily, we may fall, we may remain on the swinging bridge clinging to both knowledge and imagination in each hand. As Linton says, in relation to water generally, but it can be equally applied to groundwater:

> With respect to water and ourselves, there is no inside and no outside; water is always in our life process and so long as we live, we are always part of the water process. This statement may be taken as the hydrological imperative of a general ontology of process.
>
> Linton 2010, p. 224

For writers considering how to approach the enormous body of scientific hydrogeological knowledge, the challenge is to entangle hydrological science *with* the thinking about how inanimate entities are situated in the current philosophical discussions about human relationships to 'matter'.

As Bruno Latour and Steve Woolgar's study *Laboratory life* (1986) illustrates, it is important to distinguish between the social study of scientific method and the study of science knowledge, or the objects of the study (Latour and Woolgar 1986, p. 277). Climate writers may question the 'craft character of science', through characters who take up themes from Latour's wider questioning of how scientists create knowledge (Latour and Woolgar 1986, p. 277). For example, in the following excerpt, Frankie's conversations with Helen Mack, the hydrogeologist, turn from matters of hydrogeology to questions of the philosophy of knowledge. Frankie wants to understand how science can claim to know things without adding the contextual aspects of human and nonhuman experiences. The issue is not resolved. Helen sticks to science; Frankie continues to learn the obfuscations of knowledge silos. Hydrogeological knowledge opened a doorway to give voice to the fictionalised hydrogeologist Helen Mack in the story, and gives her verifiable information to convey. This character supports my missive to write about groundwater accurately, as closely as is currently known in science. Some readers are happy to glean facts, while others will question them (Latour and Woolgar 1986, p. 283). The monster of fallibility is always applicable, even in the most rigorous of scientific processes, and in the processes of writing groundwater narratives. Groundwater narratives explore how we might see, hear or sense groundwater's expressions through the haze of knowing and not knowing.

Excerpt 2.1: *Why we cry*, Chapter 8

Swirling weather systems whipped monsoon-like storm clouds southward and dumped sudden late-summer deluges on parched paddocks. In the Cawling district, heavy downpours picked up and washed away leaf litter, scattered branches, lifted precious soils from bare ground. It rained steadily for three days. Flash flooding spewed brackish floodwaters from the river, pushed over fences. Plastic bags and street rubbish blocked drains. Cawling residents were ill-prepared for the mess and slurry of water as it swelled from the banks of the Linden River into their streets and across the floors of low-lying homes.

Frankie called to see Helen at her office. Helen was suffering an ear infection, which gave her vertigo. She'd risked a swim in the river. The water had settled from coffee-coloured turbidity to deep tannin in the now inviting waterholes. The pools promised relief from heavy humidity, but also carried dense loads of invisible microbes, some of whom chose a home in Helen's ear canals. Now, swaying down the corridor towards her workstation, she tilted precariously towards the wall. She said the feeling made her think of the whirligigs at the local show. She looked dislocated, not sure of her direction. She shook her head, sipped on her instant coffee, straight black and thick with sugar, as Frankie now knew.

'Don't want another Hinkley here', Helen peered at a report on her computer, closed a series of screens before she turned to Frankie. Frankie knew the *Erin Brockovich* movie. She had started reading the backstory about industrial pollution of aquifers in Hinkley, California. Before she could muster a reply, Helen continued, fast, resolute.

'It's simple for me, the scientist. I'm compelled to make sense of things. It's my job to ask why. Groundwater exists in tiny and large subterranean pore spaces. Is that something strange? But.' She paused and leant towards Frankie, tipping forward on her chair. 'More importantly, what does it mean for groundwater to exist at all?'

Frankie decided to parry with another question. 'If what you thought you knew is actually wrong, namely that groundwater will last forever, if it is actually running out, how do you know that anything is right?'

'What is an aquifer, really? What's it made of? What's anything made of?' Helen persisted, leaning back in her chair to look to the ceiling, half smiling.

'Where does hydrogeology come from, does knowledge have any meaning?' Frankie swung her legs under her chair, feeling nervous at the trajectory of the tussle.

'What does any of it mean?'

They paused, took stock. The office air conditioner kicked in, a soft movement of air cooled Frankie's cheeks.

Helen continued. 'Descartes said the outside world, everything outside humans, is all measurable, mechanistically knowable. Heavy, hard and knowable.' Helen tapped at her desk. She gazed into the space between herself and the ceiling.

'Everything else then, like humans, is sensory and subjective, then.' Frankie cocked her head, birdlike.

'Yes, Cartesian thinking. Nature, culture, never to be conjoined.' Helen spread her arms wide to indicate the separation. 'Mind opposed to matter. If I can think, I exist. Nonhuman animals are machines. A dog can't feel because it hasn't got a soul. Humans alone can think and feel. You know that old line. From that, science and mathematics rule. If I don't think, I don't exist. I become a resource, open to rape and pillage by "man"-kind, of course.' She double stroked the air with two fingers to show her emphasis. 'And let's face it, that's mostly males in charge of that', Helen added, with a shrug of her mountainous shoulders.

'But if I disagree, I'm trapped. The world is nothing but ideas? There is no reality, unless it's measurable?' Frankie scratched her head, befuddled by the fast turn into philosophy. 'Don't you think that matter and the mind fit together? Humans and nature are all on the same plane.'

'Thales said that first, some dead Greek philosopher. "Everything is water."' Helen stroked the air with two fingers again to emphasise the quote. 'Imagine that. About 600 years BC. He saw us all as the one thing, connected.'

'Der.' Frankie was feeling well out of her depth, but determined to hold her own.

'It's like, now, more recently, modern physics, quantum physics and the like, is drilling down to aspects of matter that just don't follow the known rules. Things that seem to bend ideas of beingness. Sound waves, or is it light, can move through two slots, and be in both places at the same time. Some things are so big, and come and go. Like weather systems. Only computers can make funny images of them.' Helen gestured towards her computer screen. 'But how do we make sense of time and scale? Say of groundwater, for instance?' Helen had a bead of sweat forming at her temple.

Frankie shrugged, not knowing how to respond.

'Is the world as we see it really its reality?' Helen looked directly at Frankie.

'Now you're doing my head in.' Frankie paused. 'Seems to me that we make up our own realities.'

Helen pounced, hands flat onto her desk. 'Scientists don't make up anything. We only say what facts we observe, what we think we know, from research, replicable research. Anything else is', she paused, 'not yet known.'

'Things we don't know', Frankie paused, 'some call it god. There. I said the word. Not that I believe or anything. But isn't it the same thing? If you don't know something, you call on god. Mysterious.'

'Yep. Spinoza said something like that too, in the seventeenth century. Dutch philosopher. Called it pantheism. Means that god is in us too. You and me.'

'Shit. So you reckon we can only use logic, mathematics, deduction to try to understand reality. What about if you use your senses, experiences. What about imagination? Like, how I feel is a type of reality isn't it? How red are the reds I see? How purple is Kasha's nose? Isn't that just as real as some measurement?'

'You can't experience groundwater, underground', Helen persisted.

'Except maybe to drink freshly pumped bore water.'

'We know it's there from our measurements, the data we collect from drilling monitoring bores, recording the depth of the water table. The inexact data we continue to collect. We don't just believe it's there because it customarily comes from the holes we dig. The geological substrate helps us estimate the water's location, it's movements. We deduct from the data. And we stay sceptical. We question our data.' Helen was looking grittily determined.

'See, you're a scientist at core. I want to think differently about groundwater. By knowing it's there, I want to know how to preserve it. What if I'm happy to be unsure of exactly what it is, and unable to perceive it, to experience it?' Frankie paused, rubbed at her arms. 'Groundwater. All its immensity and immeasurability. Isn't it about what a thing does, more than what it means?' Frankie cocked her head, looking at the ceiling fan to find her train of thought. She continued. 'You might think that inanimate things only come to life when we lay our attention on them, our eyes, our ears, our consciousness, our caring.'

'Not like the light in the fridge? Is it really off when the door is closed?' Helen smiled.

'No, not exactly that.' Frankie flushed, trying to stay with an idea. 'It's more my concern that things have their own life, right down in their atomic activity, all the time, whether we are looking at them or not. It's the liveliness of atoms, of everything, the way we are all tangled up, connected.' Frankie paused, remembering. 'Dad used to say that steel could talk to him. I suspect that groundwater can act for itself, speak.'

'But it needs a human voice to do its representing', Helen replied, doggedly. Frankie felt herself getting tied up in knots of ideas that would not unravel.

Helen continued, 'You're talking like a philosopher. I know I invited it. But this is about geology, water in rock formations, hydrogeology.' She drained her coffee mug and pulled up a map of groundwater flow lines on her laptop.

'Seems that what we know about things has to stay in segregated compartments. Biology for this, chemistry for that, geology for that.' Frankie raised and lowered her palms, holding them apart, making separate columns in the air.

'Geology for everything.' Helen grinned. 'Some call them knowledge silos.'

'But there must be wider ways of knowing the connections.' Frankie's throat tightened. She gulped.

'Listen, I just want to show you where we think groundwater is around here. Our best estimates, our deductions. From measuring things, doing mathematics. I'll stick with the data. You stick with the fancy ideas.'

'You started it.' Frankie shook her head.

'I want to show you something.' Helen clicked on her mouse, scrolled through several menus.

Helen's boss had led a committee that developed digital maps to represent the geology and hydrogeology of the region. It brought together data sets from several government departments and agencies, which had each separately collected information about groundwater for decades.

'Complex models of groundwater's scope and depth are being aggregated on a site called Visualising Victoria's Groundwater. Some of this data has never been shared before. Public access is the important principle', Helen explained. Frankie had explored the site, fascinated by the digital elevation models of underground geological surfaces. Imagined surfaces, she thought. The mix of science, art and technology had a potency that engaged Frankie's imagination, distracting her for hours.

'OK. So it helps us imagine the invisible. But who knows how much of science is imagination anyway', Frankie persevered. Helen rolled her eyes, continued to scroll up and down, flicking at the touch pad. Coloured maps flittered as she searched for their location. She waited as data downloaded, images spread first as chequer-plate squares, then as multi-coloured landscapes that at first resembled microscopic slides of single-celled organisms. Possibilities of enmeshment between the large and the small leapt to Frankie's mind, but were beyond this conversation. Another time, she thought.

Helen began to explain the difference between saturated and unsaturated rock, the effects of geological compression on capillarity, and how water flows through rock. As often happened in their conversations, Helen turned to her notepad and began to sketch simple pictures of cross-sections of under the earth – aquifers and different rock layers represented in lines and cross-hatched spaces.

'You draw these things as if there are fixed boundaries, lines between earth and water. I know we need something, but it's like a comic. Subterranean worlds can hardly be expressed in lines.' Frankie's scepticism tumbled out as she sat back on her chair.

'This is not a Disney world we're talking about. But it's the best we have to represent water underground, its flows, its sort of shapes. Words seem to fail us.' Helen too leant back for a moment. Then, determined, turned back to her computer, continued to explain hydraulic gradient and hydraulic conductivity to Frankie. 'Different rock types allow water to flow at different rates. It also depends on the groundwater head, the height of recharge areas, and how fractured the rock is.'

'Aren't you worried that the aquifers are being wrecked?' Frankie interrupted.

'There are laws and management authorities. That's their job. I try to find the facts. That's why I work here. Now concentrate, you've got an assignment to pass.'

'OK, what was that formula again? Darcy's law.' Frankie didn't know how to explain to Helen her hope that some day groundwater would be given recognition as a legal entity, a capacity to be represented in an Australian court of law, on its own merits, its own importance. First she had to learn the science.

Expressing knowledge about/*with* a subject is affected both by the impinging resistances of material realities and by diverse cultural fluctuations in how literature makes sense of the world (Felski 2008, p. 86). Learning an array of currently understood hydrogeological facts alongside an appreciation of the uncertainty of groundwater science becomes foundational to exploring ways to include groundwater's forces and actions in narrative form. Climate writing aims to show how scientific knowledge making and affective literary knowledge making not only co-exist, but together build potentially mutually reinforcing understandings of groundwater's potency and vulnerability.

Creative Writing Research Practice

As in science research, creative practice-led research is a way to explore and create new knowledge. Curiosity about what is known and not known is a key driver. I started simply by thinking about the question, 'What is groundwater?' I learned aspects of groundwater's chemistry, biology, flows and ecological vitality through multi-layered research. My ideas developed through reading cultural scholarship, fiction, poetry and creative nonfiction, alongside science. The writing process continued to make meaning of my understandings. Based on Jamie Linton's notion that 'ideas can play an important role in the production of water – and of history – and to demonstrate that ideas do have power' (2010, p. 26), I came to see that 'water is what we make of it' through creative writing research practice (Linton 2010, p. 3). The outcomes of creative practice-led research, the creative works themselves, are unpredictable, perhaps as unpredictable as a groundwater flow through changing geologies. I mention briefly the context of creative writing practice research in universities, where the aim is to create new knowledges through creative writing research practices. Creative writing methods (the techniques writers use) and methodologies (the thinking frameworks underlying the research) are usefully distinguished in the knowledge-making process:

> In creative writing, as is the case with many other creative arts-based research disciplines, we *do*, we *make*, we *think*, in short, we *research* through practice. We discover through practice. The writing practice therefore is not just a tool for thinking but can become a form of thinking in its own right.
>
> Wardle *et al.* 2023, p. 2

Creative writing research methodologies in university settings articulate knowledge-making processes as scholarly creative writers 'make' new knowledge about how their topics are addressed in creative writing. Creative writing researchers have called the wrangle between scholarly and creative writing processes many things, for example, 'strange loops and tornadoes' (Boyd 2009) and 'nomadic emergence' (Whitehead 2013). Graeme Harper (2013) calls the writing process 'responsive critical understanding', and suggests that the key is the 'fluidity of the relationship between creative work and critical response' (p. 285). Kim Lasky utilises the concept of 'poetics' to articulate the hinges within the 'triptych of practice-led research' (2013, pp. 21–22). Poetics, Lasky (2013, pp. 21–22) argues, involves reflexive consideration of the many influences within the other two 'wings' of the triptych: research for the creative work on one side and the critical and theoretical inputs and outcomes on the other. Lasky's model gives attention to the two-way process between 'poetic' reflections on the creative work and the scholarly inputs which enhance each writing form. My creative writing research practice begins with creative writing and responds in a cyclic manner with my critical responses to my scientific and scholarly learning. As Harper (2013) suggests, creative writing research starts with the writing practice. For example, as I wrote scenes towards *What we cry* during my PhD, I concurrently learnt foundations of the science of hydrogeology. The indeterminacy of groundwater and hydrogeology became evident. This led me to read philosophy of indeterminate entities, such as Jane Bennett's *Vibrant matter: a political ecology of things* (2010), which led me back to the novel-writing process with a new critical appraisal of how I was attempting to convey ideas about groundwater. In these ways creative writing research opens possibilities for multiple interactions and influences between the background research for the content of the creative climate writing, particularly cultural and literary theory, environmental philosophy, science of hydrogeology, water politics and water management. At the same time, the creative writing process itself feeds the development of new critical knowledge and new research ideas (Arnold 2012; Holloway 2013; Kroll 2013).

In the research for groundwater narratives, I call the spiralling interactions *oscillating absorption*. Both oscillation and absorption occur between creative writing practice, critical theoretical research, and reading, observing, reflecting and thinking. In a scholarly context, a researcher might start a research process by presuming that research as a creative writer-as-researcher is merely 'the physical act of inscribing words on paper', including the 'drafting and redrafting until the final manuscript is arrived at' (Brien 2006, p. 54). Writing self reflective groundwater narratives as part of academic scholarship becomes more than this. It includes observing and articulating the processes by which the writer 'absorbs' the various influences on their writing. Through *oscillating absorption* between reading and writing forms, expressions of groundwater's presence and agency emerge as the primary subject of the writing process. Entanglements between art practice and art research become so close as to at times appear contrived (Borgdorff 2011, p. 45). Barrett and Bolt argue that the 'generative capacity of creative arts research is derived from the subjective, emergent and interdisciplinary approaches' (2010, p. 3). Groundwater narratives are distinctively interdisciplinary, as writers oscillate between hydrogeology, cultural and literary theory and creative writing practice. Writing into the spaces of new knowledge is inherent to *oscillating absorption* as a creative writing research method and methodology.

In summary, a combination of the effects of global warming and the expanding forms of climate writing continues to expand the conventions of climate writing into posthuman, new materialist ways of knowing matter. Writing groundwater narratives in the Anthropocene may not completely shuck the anthropocentric tendency of human language – even as it attempts to emphasise the value of thinking beneath the surface – but the potential remains for subterranean imaginaries to make significant contributions to climate writing that expresses the potency of matter which exists under the water table and is not at first seen.

Community and scientific knowledge about global warming is changing fast. Urgency about the consequences of climate change continues to amplify while human responses remain perilously partial and fragmented (Hamilton 2017; White and Nelson 2018; Wiseman 2021). Climate writers write into a rapidly transforming time, where groundwater is also increasingly following a trajectory of its own. I next consider how ascribing meaning to groundwater's narrative potency utilises creative writing processes of storying *with*. Storying *with* groundwater invites climate writers into territory where what we know and do not know about groundwater's materiality, its relationality and its agency really matters.

Notes

1 Key thinkers in this terrain include new materialists and material ecocritics (Coole and Frost 2010; Iovino 2014), posthumanists (Braidotti 2013), non-representationalists (Vannini 2015) and eco-feminists (Rose 2013; Neimanis 2017; Plumwood 2018; Rigby 2018; Stevens, Varney and Tait 2018), to mention just some of the relevant scholars that have influenced how writers might know, imagine and express the significance of matter. Narrative groundwater takes its place in this body of literature.

2 Luce Irigaray's *This sex that is not one* is quoted in Yvonne Volkart's (2022) inspiring essay, 'Counter appropriation: flow, flood, fib'. Irigaray wrote with the concepts of fluidity and water, saying 'The fluid – as that other, inside/outside of philosophical discourse – is inherently unstable. Unless one subjects it to geometrisation or (?) idealizes it' (quoted in Volkart 2022, p. 2, question mark in the original).

3 Parts of this section first appeared in an essay of mine in the Canadian literary journal *Mosaic* (Wardle 2019).

4 See an example of a landscape cross-section, surface waters and groundwater zone at MicrobeWiki (2011).

References

Alaimo, S. and Hekman, S., 2008a. Introduction: emerging models of materiality in feminist theory. *In:* S. Alaimo and S. Hekman, eds. *Material feminisms*. Bloomington, IN: Indiana University Press, 1–19.

Alaimo, S. and Hekman, S., eds, 2008b. *Material feminisms*. Bloomington, IN: Indiana University Press.

Álvarez-Miranda, B.F., 2021. When matter takes a position: post-anthropocentric landscapes in contemporary art. *In:* D. Villanueva-Romero *et al.*, eds.

Imaginative ecologies: inspiring change through the humanities. Leiden: Brill, 137–156.

Arnold, J., 2012. Practice led research: creative activity, academic debate, and intellectual rigour. *Higher Education Studies*, 2 [online]. Available from: http://dx.doi.org/10.5539/hes.v2n2p9 [accessed 19 January 2023].

Bakhtin, M.M., 1981. *The dialogic imagination: four essays.* Austin, TX: University of Texas Press.

Barad, K., 2007. *Meeting the universe halfway: quantum physics and the entanglement of matter and meaning.* Durham, NC: Duke University Press.

Barad, K., 2008. Posthumanist performativity: towards an understanding of how matter comes to matter. *In:* S. Alaimo and S. Hekman, eds. *Material feminisms.* Bloomington, IN: Indiana University Press, 120–154.

Barad, K., 2015. Transmaterialities: trans*/matter/realities and queer political imaginings. *GLQ: A Journal of Lesbian and Gay Studies*, 21, 387–422.

Barad, K., 2017. No small matter: mushroom clouds, ecologies of nothingness, and strange topologies of spacetimemattering. *In:* A.L. Tsing *et al.*, eds. *Arts of living on a damaged planet: ghosts and monsters of the Anthropocene.* Minneapolis, MN: University of Minnesota Press, 103–120.

Barrett, E. and Bolt, B., 2010. *Practice as research: approaches to creative arts enquiry.* London: I.B. Tauris & Co.

Baucom, I. and Omelsky, M., 2017. Knowledge in the age of climate change. *South Atlantic Quarterly*, 116, 1–18.

Bennett, J., 2010. *Vibrant matter: a political ecology of things.* Durham, NC: Duke University Press.

Borgdorff, H., 2011. The production of knowledge in artistic research. *In:* M. Biggs and H. Karlsson, eds. *The Routledge companion to research in the arts.* Abingdon: Routledge, 44–63.

Boulter, A., 2007. *Writing fiction: creative and critical approaches.* Basingstoke: Palgrave Macmillan.

Boyd, N., 2009. A creative writing research methodology: new directions, strange loops and tornados. *In:* D.L. Brien and M. Freiman, eds. *Margins and mainstreams: refereed conference papers of the 14th Annual AAWP Conference.* Hamilton, NZ: Waikato Institute of Technology.

Braidotti, R., 2013. *Posthuman.* Oxford: Wiley.

Brien, D.L., 2006. Creative practice as research: a creative writing case study. *Media International Australia*, 118 (1), 53–59.

Buell, L., 2001. *Writing for an endangered world: literature, culture, and environment in the U.S. and beyond.* Cambridge, MA: Belknap Press of Harvard University Press.

Buell, L., Heise, U. and Thornber, K., 2011. Literature and environment. *Annual Review of Environment and Resource*, 36, 417–440.

Chen, C., Macleod, J. and Neimanis, A. 2013. Introduction: towards a hydrological turn? *In:* C. Chen, J. Macleod and A. Neimanis, eds. *Thinking with water.* Montreal, QC: McGill-Queen's University Press, 3–22.

Clarke, B. and Rossini, M., 2011. *The Routledge companion to literature and science.* London: Routledge.

Coole, D. and Frost, S., 2010. Introducing the new materialisms. *In:* D. Coole and S. Frost, eds. *New materialisms: ontology, agency and politics.* Durham, NC: Duke University Press, 1–43.

Currell, M., Gleeson, T. and Dahlhaus, P., 2016. A new assessment framework for transience in hydrogeological systems. *Groundwater*, 54, 4–14.

Daley, L., 2014. Rendering visible: painting and sexuate subjectivity. *Educational Philosophy and Theory*, 47, 608–621.

Deleuze, G. and Guattari, F., 1988. *A thousand plateaus: capitalism and schizophrenia.* London: Athlone Press.

Felski, R., 2008. *The uses of literature.* Malden, MA: Blackwell Publishing.

Fitts, C.R., 2013. *Groundwater science.* Waltham, MA: Academic Press.

Glotfelty, C., 1996. Introduction: literary studies in an age of environmental crisis. *In:* C. Glotfelty and H. Fromm, eds. *The ecocriticism reader: landmarks in literary ecology.* Athens, GA: University of Georgia Press, xv–xxxvii.

Guattari, F., 2000. *The three ecologies.* London: Athlone Press.

Hamilton, C., 2017. *Defiant Earth: the fate of humans in the Anthropocene.* Sydney: Allen and Unwin.

Haraway, D., 2004. *The Haraway reader.* New York: Routledge.

Haraway, D., 2016. *Staying with the trouble: making kin in the Cthulucene.* Durham, NC: Duke University Press.

Harper, G., 2013. Creative writing research. *In:* G. Harper, ed. *A companion to creative writing.* Hoboken. NJ: John Wiley & Sons, 278–290.

Herbrechter, S. and Callus, I., 2008. What is a posthumanist reading? *Angelaki: Journal of Theoretical Humanities*, 13, 95–111.

Holloway, S., 2013. How to make a pocket watch: the British PhD in creative writing. *In:* G. Harper, ed. *A companion to creative writing.* Hoboken, NJ: John Wiley & Sons, 129–143.

Iovino, S., 2014. Bodies of Naples: stories, matter, and the landscapes of porosity. *In:* S. Iovino and S, Oppermann, eds. *Material ecocriticism.* Bloomington, IN: Indiana University Press, 97–113.

Iovino, S. and Oppermann, S., 2014. Introduction: stories come to matter. *In:* S. Iovino and S. Oppermann, eds. *Material ecocriticism.* Bloomington, IN: Indiana University Press, 1–17.

Jagodzinski, J., 2021. Quantum creativity: afracting new materialism in the Anthropocene. *Qualitative Inquiry*, 28, 586–595.

Kostkowska, J., 2013. *Ecocriticism and women writers: environmentalist poetics of Virginia Woolf, Jeanette Winterson and Ali Smith*. New York: Palgrave Macmillan.

Kroll, J., 2013. The creative writing laboratory and its pedagogy. *In:* J. Kroll and G. Harper, eds. *Research methods in creative writing*. Basingstoke: Palgrave Macmillan, 102–132.

Lasky, K., 2013. Poetics and creative writing research. *In:* J. Kroll and G. Harper, eds. *Research methods in creative writing*. Basingstoke: Palgrave Macmillan, 14–33.

Latour, B. and Woolgar, S., 1986. *Laboratory life: the construction of scientific facts*. Princeton, NJ: Princeton University Press.

Le, V., 2017. Schelling and the sixth extinction: the environmental ethics behind Schelling's anthropomorphization of nature. *Cosmos and History: The Journal of Natural and Social Philosophy*, 13 (3), 107–129.

Linton, J., 2010. *What is water? The history of a modern abstraction*. Vancouver: UBC Press.

Linton, J. and Budds, J., 2014. The hydrosocial cycle: defining and mobilizing a relational-dialectical approach to water. *Geoforum*, 57, 170–180.

Massumi, B., 2001. Introduction: like a thought. *In:* B. Massumi, ed. *A shock to thought: expression after Deleuze and Guattari*. London: Routledge, xiii–xxxix.

Mathews, F., 2021. *The ecological self*. London: Routledge.

MicrobeWiki, 2011. *Groundwater* [online]. Available from: https://microbewiki.kenyon.edu/index.php/Groundwater [accessed 20 January 2023].

Morton, T., 2007. *Ecology without nature: rethinking environmental aesthetics*. Cambridge, MA: Harvard University Press.

Morton, T., 2013a. *Hyperobjects: philosophy and ecology after the end of the world*. Minneapolis, MN: University of Minnesota Press.

Morton, T., 2013b. Poisoned ground: art and philosophy in the time of hyperobjects. *Symploke*, 21, 37–50.

Neimanis, A., 2017. *Bodies of water: posthuman feminist phenomenology*. London: Bloomsbury Academic.

Plumwood, V., 2008. Shadow places and the politics of dwelling. *Australian Humanities Review*, 44, 139–150.

Plumwood, V., 2009. Nature in the active voice. *Australian Humanities Review*, 46, 113–129.

Plumwood, V., 2018. Ecofeminist analysis and the culture of ecological denial. *In:* L. Stevens, P. Tait and D. Varney, eds. *Feminist ecologies: changing environments in the Anthropocene*. Gewerbestrasse, Switzerland: Palgrave Macmillan, 97–112.

Rigby, K., 2018. Women and nature revisited: ecofeminist reconfigurations of an old association. *In:* L. Stevens, P. Tait and D. Varney, eds. *Feminist ecologies: changing environments in the Anthropocene.* Gewerbestrasse, Switzerland: Palgrave Macmillan, 57–82.

Rose, D.B., 2013. Slowly – writing into the Anthropocene. *TEXT,* Special Issue 20.

Rose, D.B., 2014. Arts of flow: poetics of 'fit' in Aboriginal Australia. *Dialectical Anthropology,* 38, 431–445.

Segal, M., 2017. The missing climate change narrative. *South Atlantic Quarterly,* 116, 121–128.

Sleigh, C., 2011. *Literature and science.* Basingstoke: Palgrave Macmillan.

Stevens, L., Tait, P. and Varney, D., 2018. Introduction: 'street fighters and philosophers': traversing ecofeminisms. *In:* L. Stevens, P. Tait and D. Varney, eds. *Feminist ecologies: changing environments in the Anthropocene.* Gewerbestrasse, Switzerland: Palgrave Macmillan, 1–22.

Stevens, L., Varney, D. and Tait, P., eds., 2018. *Feminist ecologies: changing environments in the Anthropocene.* Gewerbestrasse, Switzerland: Palgrave Macmillan.

Thomashow, M., 2002. *Bringing the biosphere home: learning to perceive global environmental change.* Cambridge, MA: MIT Press.

Van Loon, J., 2023. Not-knowing. *In:* D. Wardle *et al.*, eds. *A to Z of creative writing methods.* London: Bloomsbury Academic, 114–116.

Vannini, P., 2015. Non-representational research methodologies: an introduction. *In:* P. Vannini, ed. *Non-representational methodologies: re-envisioning research.* New York: Routledge, 1–18.

Volkart, Y., 2022. Counter appropriation: flow, flood, fib. *Insert,* 2 [online]. Available from: https://insert.art/ausgaben/senseabilities/counter-appropriation/ [accessed 29 January 2023].

Wardle, D., 2019. Groundwater as hyperobject. *Mosaic: An Interdisciplinary Critical Journal,* 52 (2), 1–16.

Wardle, D., *et al.*, 2023. Introduction. *In:* D. Wardle *et al.*, eds. *A to Z of creative writing methods.* London: Bloomsbury Academic, 1–6.

White, E.K. and Nelson, R.L., 2018. What happens to small towns whose water becomes big business for bottled brands? *The Conversation,* 17 May [online]. Available from: www.abc.net.au/news/2018-05-17/what-happens-to-small-towns-whose-water-becomes-big-business-fo/9770128 [accessed 5 July 2018].

Whitehead, H., 2013. Nomadic emergence: creative writing theory and practice-led research. *In:* G. Harper, ed. *New ideas in the writing arts.* Newcastle upon Tyne: Cambridge Scholars Publishing, 95–112.

Wiseman, J., 2021. *Hope and courage in the climate crisis: wisdom and action in the long emergency.* Cham, Switzerland: Springer International Publishing.

3

SPRINGS AND SEEPS

STORYING *WITH*

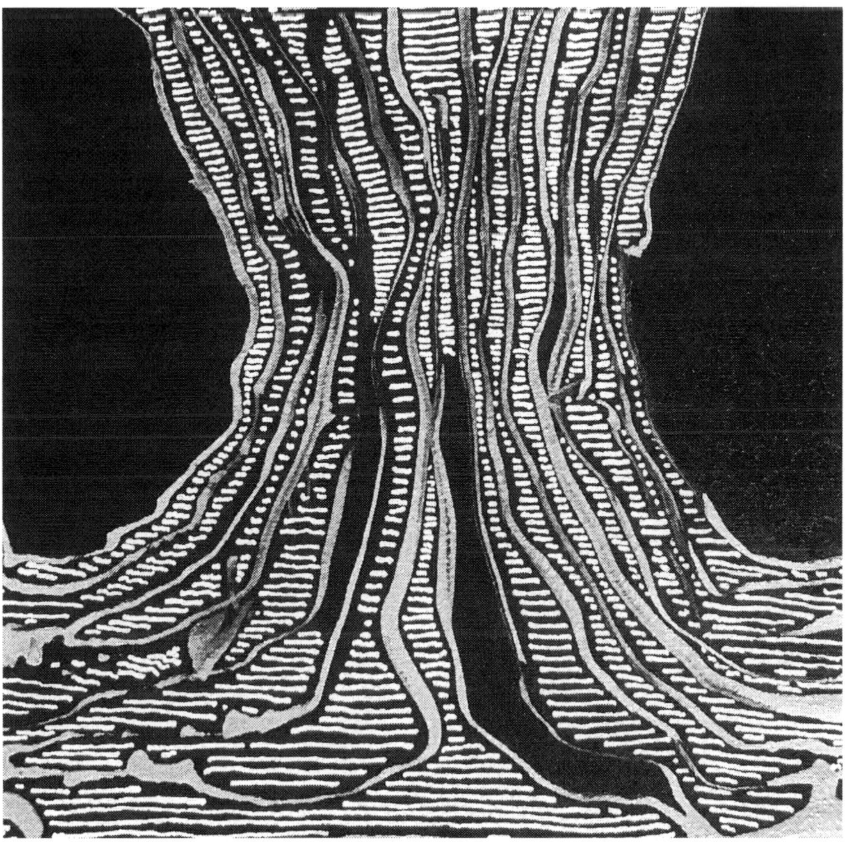

Figure 3.1 Bill Sampson, *Drawing water*, 2022, synthetic polymer and enamel on aluminium as digital photograph, 11 × 11 cm.

DOI: 10.4324/9781003270416-3

Springs are where groundwater discharges to the surface. Where percolating groundwater moves downward and reaches the water table, or an impermeable layer of rock, it flows laterally till it meets the surface as a spring or a slow seep. Springs and seeps commonly occur at the base of hills and mountains. Springs respond to different geologic formations. For example, they may occur where permeable sands meet impermeable slates or shales; they may emerge through fractured igneous rocks where the water table is high. Springs emerge where limestone cavities carrying water meet either the surface or impervious rock layers. Springs might also seep as base flow into the beds of rivers, streams and creeks (permanent and ephemeral). This occurs when the elevation of the water table in the surrounding landscape becomes higher than the riverbed. Springs may also flow unseen into lake beds and ocean beds. The source of spring water may be local or far distant. Spring water may have been subterranean for short periods (weeks or days) or for thousands of years before it reaches the surface. The chemistry and temperature of springs varies according to the composition of the rocks they have passed through, the time the water takes between recharging and emergence, and the depths from which they emerge. Hot springs come from water's contact with geological heated rock formations deep below the surface. They may emerge at boiling temperature.

> [T]he novelist writes amidst the bounds and liberations of the logic of storytelling.
>
> Sontag 2007, p. 216

Stories flow. Water flows. The idea of flow implies changes and becoming something new. As Sontag guides us, storytelling provides both the limits and freedoms of water-like flow. Storying with groundwater too has its 'bounds and liberations'. Groundwater's capacity to push to the surface, responding to the geologies through which it passes, works as an analogy for storying *with* principles and processes, which will be explored in this chapter.

Australia, the driest inhabited continent on the planet, is scattered with springs, sometime in unlikely desert terrain. Spring waters have held great interest for eons as they ostensibly appear from 'nowhere', bringing water's salve, its healing capacities, and at times springs are connected to mystical beliefs. The sources of springs may be unknown to many settler community members, but can often be traced to far-distant groundwater flows either through Indigenous knowledges (Mahood 2022) or by using hydrogeological knowledge of deep geologic formations. Springs have been tapped and contained for centuries, and have been the centre point around which communities and cultures evolve. From Bedouin villages to Mongolian high mountain trading routes, to Indigenous Australian songlines along what is now called the Oodnadatta Track, and across the Australian continent, people collect around springs. Stories flow. Water flows. The idea of flow implies changes and becoming something new. Climate change discourses are reshaping intellectual and artistic disciplines, and the narratives within and across humanities and science (Baucom and Omelsky 2017, p. 2).

Storying *with* gives attention to relational ways of meaning making (Phillips and Bunda 2018, pp. 56–61; Bunda *et al.* 2019). Groundwater narratives utilising storying *with* principles create meaning in the written form through unfolding processes of writing-reading-writing groundwater's viscosity, its flows and enmeshments. Groundwater narratives express the stickiness of groundwater's entanglements with multiple critters, microorganisms, stygofauna, with forests and clouds, and with the threats to its free-flowing, cyclic existence. Storying *with* groundwater means weaving the various threads of human and nonhuman lives with groundwater in many different textures and flows. As readers/writers turn towards the potency of the real and material impacts of anthropogenic climate change, a consideration of critically storied human enmeshments with the nonhuman world becomes an essential writerly response.

Through storying *with*, groundwater becomes something more than imaginary. Though humans may not usually see groundwater in its geological porous places, it is something real. Through storying *with*, its reality and its impacts are revealed. Chen *et al.* (2013) provide foundational approaches for 'thinking *with* water', explaining how human and nonhuman entanglements with water's lively materiality provide the essential '*with*' that underpins climate writing processes. Storying *with* is 'lively and emergent', enabling intertwined and interactional intricacies between the storyteller and their subjects (Phillips and Bunda 2018, p. 73). Storying *with* contributions from subterranean imaginaries draws attention to the 'primacy of storying and making and for provoking resonant and entangled understandings of belonging and displacement' (Bunda *et al.* 2019, p. 155).[1] Writing groundwater narratives entails many lively entanglements between the writer and the aquifers and springs to which they are connected.

The principles 'gathered' by Louise Phillips and Tracey Bunda (2018, p. 7) to describe storying as 'making and remaking meaning through stories' are primarily connected to the oral traditions of Indigenous Australians. These five principles of storying are proposed here as part of a repertoire for climate writers because of their relevance to the challenges of storying in the Anthropocene:

> 1 storying nourishes thought, body and soul;
> 2 storying claims voice in the silenced margins;
> 3 storying is embodied, relational meaning making;
> 4 storying intersects the past and present as living oral archives; and
> 5 storying enacts collective ownership and authorship.
>
> Phillips and Bunda 2018, p. 45

Storying as a research and writing method has a strong tradition in Indigenous art making and research, and forms a significant strategy of resistance to white colonial perspectives (Todd 2015; Phillips and Bunda 2018, pp. 4–8). Acknowledging First Nations roots in storying principles is essential in maintaining ongoing respect for Indigenous voices. First Nations writers and storytellers in Australia and other colonised countries such as the USA, Canada and New Zealand are reclaiming lost languages and millennium-old oral storytelling traditions that reconnect Indigenous custodians to Country.[2] Aboriginal storying, as proposed by Phillips and Bunda (2018, p. 43), roots stories and storying with Country, with cultural places, and from these stories cultural meaning is made. The power of oral traditions lies in connecting ancient cultural lore with contemporary experience. Indigenous knowledge of ageless connections with lands and waterways comprehends narrative groundwater's capacity to cross time-frames. For narrative groundwater, immediate presence and ancient pasts intermingle. Both oral storytelling and Indigenous authors' written texts make meaning of Aboriginal and Torres Straight Islanders' experiences of genocide, displacement and longstanding resistance to colonial theft of land, language and children. The intended potency of storying comes from its depth in Indigenous culture and in the specific context of reclaiming stories in the context of loss and displacement of First Nations peoples since colonisation.

It is vitally important to acknowledge that Indigenous storying, with its roots in an oral tradition, cannot be replicated in white settler climate writing. There are limits to white writers' understandings of Indigenous contexts of storying, which risks an incomplete use of the complex ideas and experiences that are woven beneath the surface of Phillips and Bunda's five principles of storying. Speaking about colonised peoples from ill-informed, white settler perspectives risks insult and appropriation of Indigenous stories (Wright 2016). Phillips and Bunda's proposal that storying 'is inquiry, is theorising, is sharing/presenting of research' (2018, p. 5) has wide ethical resonance.

Storying processes can be applied by climate writers to storying *with* groundwater, Anglo/settler and Indigenous writers alike. In groundwater narratives, the term narrative predominantly relates to the written form of language that embeds and embodies human and nonhuman characters in configured plots. The five principles are relevant to climate writers who predominantly give written expression to groundwater's many ways of becoming, but they must be applied with sensitivity, respect and deep listening to First Nations' relationships to Country and waters. In an Australian context, the *Protocols for using First Nations cultural and intellectual property in the arts* spells out the legal, ethical and moral considerations for Indigenous and non-Indigenous writers and artists using First Nations cultural material (Janke 2019, p. 1). As highlighted in Chapter 1, these are critical processes to understand and practise in storying. For instance, use of First Nations language in any projects requires the free, prior informed consent of custodians of that language. For example, in *Why we cry*, through respectful written and oral communications with the local Indigenous cooperative, Dja Dja Wurrung Cooperative, I obtained formal permission to use a local Indigenous word, *yaluk*, for the name of the fictional organisation in the novel – Yaluk Cooperative. The word *yaluk* means creek or river.

Deep listening to Indigenous stories is a slow process that enables Indigenous storytellers and both settler and Indigenous listeners to find their places in the narratives of dispossession, and to begin the work of decolonisation (McGraw and Pieris 2015, pp. 19–20). Climate writers can learn from reading and listening to Australian Indigenous writers, ranging from nineteenth-century activist, writer and inventor David Unaipon, to twentieth-century poet Oodgeroo Noonuccal, to many contemporary scholars and writers who have used the written word to advocate for rights and social justice. Deep listening includes respecting Indigenous forms of storying. Amanda Johnson highlights the work of selected First Nations Australian poets to exemplify the rich genealogy of ecopoetics that:

> deploy intertextual strategies, surreal imaging, polyphonic linguistic mashups, cinematic disjuncture, polyphonic dialogism, hyper-anthropomorphism, radically inventive rhyme schemes, gallows humour and concrete language strategies (and *language* strategies). While some of these techniques suggest dedicated flirtations with postmodern poetic mechanics, such formal and conceptual complexity is distinctive to modes of Aboriginal realism that interweave stories of place and spiritual belief with political struggle and resistance under colonialism.
>
> Johnson 2021, p. 19

Recent voices in the strong First Nations chorus include: Professor Megan Davis (2020) for her contributions to the creation and recognition of the Uluru Statement from the Heart; Jeanine Leane (2016), whose poems and essays call for Indigenous voice; Alexis Wright (2016), whose prize-winning fiction and essays continue to speak of the deceit and injustice of appropriating Indigenous stories; Bruce Pascoe who, as well as his best-selling book *Dark emu* (2014), also regularly writes on the consequences of historic and ongoing colonisation (Pascoe 2020); award-winning fiction writer, essayist and historian Tony Birch (2016, 2020), who has long championed the importance of Indigenous voices in climate change narratives; and young and queer poet, fiction writer and Mununjali (Yugambeh language group) educator Ellen Van Neerveen (2019), who speaks and writes of ongoing colonial injustices, alongside their award-winning creative writing that illuminates Indigenous voices. These and many other Indigenous writers and scholars consciously claim the power of storytelling as a way to resist and refuse ongoing colonial actions.[3]

Oral traditions of storytelling are highly significant in many First Nations cultures and follow complex and ageless themes. Indigenous storytellers have for millennia included an appreciation of groundwater, 'characterising it through contextual layers of creation stories', which remain paramount in contemporary Aboriginal law and legal obligations (Marshall 2017, p. 21). A strong example of Australian Aboriginal connection to groundwater is portrayed orally in the documentary *Putuparri and the rainmakers* (Ma 2015). Tom 'Putuparri' Lawford and his family occupy Country in the Western Desert, near Fitzroy Crossing, Western Australia. Tom's grandfather, 'Spider Putuparri' was one of the last remaining rainmakers who passed on cultural stories of Kurtal, the Country their family had occupied for tens of thousands of years. He took Tom as a young boy to Kurtal, including to a highly significant waterhole in the Great Sandy Desert, telling him the stories of the place. When cared for by custodians it is the site where artesian water, called '*djili*' or 'living water', comes to the surface and where the spirits of Putuparri's people return to over the millennia. When Tom and community members return to the neglected spring, blocked probably by cattle, on his grandfather's instructions they dig sand to clear the spring. Spider's calls to the spirits of the spring are not 'imaginary'. He calls out to the animating spirit of the water, entering into dialogue with it. The response from Country is remarkable. Finally the rain comes.

The hydrosphere is under attack as water is increasingly ignored, commodified, polluted and taken from parts of the global water cycle in ways that cannot be replenished. Thinking *about* water's problems across the global water crisis has predominantly cast groundwater as a minor actant. Thinking *about* water implies our separation from water, substantially giving us a licence to think of water as a resource that needs to be managed and used. Access to water may at first be considered either as a common public and cultural right, or as something that can be commodified and purchased according to market principles. Wary of false dichotomies, there are usually more complex arrays of positions on access to water than public versus private rights (Bakker 2010). Bakker (2010, p. 158) points out that there are potentially unknown environmental and social consequences of simply advocating for unregulated human rights to water. Many books and films position stakeholders around these conflicting perspectives, addressing various ways communities might manage water scarcity and oppose appropriation and pollution of water sources (Chen *et al.* 2013, p. 3). *Putuparri and the rainmakers* (2015) is one of the few that give voice to Indigenous custodianship of water. How governments negotiate political, economic and social conflicts over groundwater and surface water rights systems, usually without the wisdom of Indigenous custodians, are just some of the issues that might be addressed in groundwater narratives. Storying *with* groundwater gives greater precedence to groundwater's enmeshments in these narratives, as troubled as these enmeshments may be.

Encapsulating all of water's forms and complex relations is virtually impossible. Many cultural, social and political forces impinge on many of water's expressions. Stories bubble and percolate to the surface consciousness of climate writers in both spontaneous and hard-laboured ways. Stories are as life-giving as springs. Writers responding to the Anthropocene produce stories with the richness that a spring brings to a parched landscape. Expressing tragic desiccation that follows the drying up of spring-fed water sources may also be part of a story. Like springs, stories emerge in as many ways as there are ways to breathe. They are influenced by the internal palimpsest that writers bring to their work; the layerings of heart, passions and insights into particular subjects, the workings of intellect and memory, the histories and political contexts, the social and cultural norms that scrape at the author. The list goes on. Bakhtin sees these prerequisites to writing as crucial for building a dialogical text, one which is a reflection of complex 'social heteroglossia' (1981, p. 264). The variety of 'voices' that contribute to authentic authorial expression gives a sense of palimpsest to the tasks of storying *with* groundwater. Palimpsest comes from the Greek *palimpsēstos*, meaning scraped again, originally referring to a page of manuscript that could be scraped clean then used again. In geology, the term palimpsest refers to an exposed rock surface reveals geological processes from different ages. Timeless actions of geology are scraped into a rock surface – sliding rock layers in a fault, a grinding glacier, a volcanic slurry passing by.

Writing groundwater narratives involves the co-imbrication of temporality and spatiality; scraping pasts, writing, rewriting, exposing over time new and old 'springs' from our subterranean imaginaries. The palimpsest is layered with the voices of groundwater itself as it percolates and seeps. The text is ever developing, always changing as words, paragraphs and stanzas are erased to make room for revised text. Narratives, plot lines and characters' qualities develop alongside the author's thinking about what groundwater means and what it does. Different contexts of groundwater's actions and emergences in springs materialise.

Water means vastly different things in different contexts. Various Indigenous custodians of lands and waters may think of water in divergent ways to white settler water managers who may hold a colonising, managerial view of the water systems they attempt to manipulate, usually for economic gain (Moggridge 2010; Marshall 2017). The task of climate writers is to complicate these viewpoints, to explore the experiences that lie within the imaginaries of all players, including that of groundwater itself.

Thinking about the processes of storying *with* groundwater in narrative long-form fiction includes all the skills and techniques of novel writing: character development, coherent plotting, clear timeframes, relatable settings, consistent voices, innovative structuring, clarity of language, use of metaphor and mood. Storying *with* involves imaginative techniques that derive from the plethora of writing techniques that climate writers consciously choose to adopt. Textbooks about the steps and stairs of novel writing skills abound.[4] Storying *with* groundwater is enmeshed with each of these elements of writing climate fiction.

How might storying *with* adopt a narrative style that has both contemporary political relevance and an aesthetic form to engage readers? Nick Admussen's (2016) six proposed reforms to literature in the Anthropocene provide clues:

1. Reject progress narratives.
2. Retire the portrait of the single soul.
3. Stop yelling over biomes.
4. Acknowledge the blight of global inequity and material poverty of people and nations exploited by industrialisation.
5. Choose systems of membership and interrelation over objects and metaphors of possession and exchange.

6. Transform the relations between imagination and actions: 'literature can no longer hang outside the world' (p. 6).

Admussen proposes that climate writers globally have an ethical responsibility to build new forms of literature that respond to the challenges of climate change. Storying *with* groundwater's narrative capacity juxtaposes tenably with Admussen's rallying call to tell the stories that matter.

Storying *with* in the Anthropocene also means writing slowly and ethically, writing to witness the unquantifiable damage humans are causing, as well as writing for the hope and tenacity of both humans and more-than-human entities (Rigby 2009; Rose 2013). Climate writers' creative task is to witness, to hear the possibilities for metaphor and analogy and to pay attention to the story-making qualities of groundwater. How groundwater might be known and how it is affected by humans, as well as how groundwater creates affects in all its enmeshments, is the underlying force to groundwater narratives. In the information-rich, increasingly 'post-truth' world, where facts are contrived to suit the moment, it is increasingly important to consider the ethics of writing. I seek not to contribute to the stories that centre and normalise commercialisation of groundwater. Groundwater narratives embrace the 'ethical potential of narrative to communicate the incommunicable … [and] of fiction as storytelling as a way to build a shared terrain of experience and to rebuild our possibilities for exchange' (Tynan 2020, p. 47). Climate writers aim to narrate worlds that bend or diffract readers' thinking away from capitalist assumptions of commercialisation and towards a reverence for sustainable subterranean aquifers.

Metaphors

My growing understandings of aquifers as potent underground entities enable me to explore meaning and metaphor in their presence and absence. Engaging with metaphors opens possibilities for avoiding false hopes of mimetic expressions in subterranean imaginaries (Felski 2008, pp. 82–85). A metaphor asserts a similarity between two or more different things (Barnes and Duncan 1992, p. 10). A distinction between analogy and metaphor is useful: in analogy (simile) the 'this *is like* that' works as a means of clarification, whereas the metaphorical 'this *is* that' operates as a means of subversion or expansion of ideas (MacFarlane 2013, pp.

104–105). Metaphors can create new knowledge. Climate writers' creative tasks include to witness, to pay attention and hear the possibilities for metaphor and analogies, the story-making qualities of groundwater. Within the limits of fabricating an allegorical world, storying *with* utilises expressions of a recognisable world to expand the voices of material, nonhuman entities. Writing about geology leads me to imaginary places where metaphors assist making meaning. Metaphoric evocations build an imaginary world for readers. Narrative groundwater necessarily takes climate writers to the realms of metaphor.

Imagining groundwater's place in the water cycle, or rather the hydrosocial cycle, is to see fluids moving from internal to external, back to internal. Cleansing. Life giving. Exploring the potency of metaphors and metonyms in groundwater narratives may evoke ideas of groundwater as lubricator of 'mother earth'. Imagine the planet needing moisture under her skin, storing and expelling water from her orifices. Liquids are held internally, until, like sweat through skin pores, like pee through a urethra, like tears from the lacrimal gland, they emerge at the surface. Groundwater could be seen like blood and salt in veins, fluids in the lymphatic system, moving to nodes, carrying enzymes and hormones. Groundwater pools beneath the surface in aquifers, like lymphoedema, stretching skin. Water, oils and gases bubble and squelch through and within layers, crevices and pores of the crust. Saturation is when all the spaces between the particles of stone and rock are filled, and no air remains. There is something breathless in that.

Groundwater's pulse comes from grand geological breathing, the gravitational dance between elastic earth and moon. Geological forces puff volcanoes to the surface with magnitude-defying powers. Movements of tectonic plates exert pressure on permeable rock, till it warps into mountain ranges like crimps in a gripped sponge. Water flows and steams, hisses and flies in heat. A visit to any hot water spring in deserts or icy mountains, or to the spouting geysers at Rotorua, can only inspire the potency of water meeting with internal geological forces.

Potential for analogies abound in the ways that contamination plumes through aquifers. Plume is the word used by hydrogeologists for pollutants spreading by dispersion (mechanical) and diffusion (molecular movement through a fluid). A plume may be a feather, a tuft of feathers on a bird's tail. A plume, a flow of matter, especially of waste material or pollutant, spreading from a source. A plume is also an ornament or token of honour.

A single word can bounce and diffract different meanings. Groundwater's dispersive flow might act as a metaphor for a protagonist's life, for human and nonhuman sojourns to unchartered places. Groundwater may be a metaphor for the interior life of humans, for the tears we shed. We are comprised of waters, within our skin, our muscles, our bones, our heart, our brain. The potential for figurative mapping of groundwater is limited only by a writer's imagination. Metaphors or allegories for groundwater easily become unstuck, slippery and elusive if the rhetorical use of language is too vague (Haser 2005, pp. 240–241). Climate writers utilise them with care, understanding that there may be many different conceptions of the terms of the metaphor – both the target and the source terms (Haser 2005, p. 244).[5]

Writers are bound to look to the contexts of words for the meaning and imaginaries they may convey. Gaston Bachelard's concept of 'intimate immensity' in *The poetics of space* helps us to consider ways to transcend 'the contradiction of small and large' (1994, p. 190). Thinking *with* groundwater involves many instances of moving from the microscopic to the vast, from cellular to immense water bodies. Juxtaposing intimacy and immensity, Bachelard brings together, through poetic imagining, the powers of imaginary spaces (Danielewski 2014). Bachelard's exposition on the rapport between imagination and language inspires the notion that images may 'speak' the emergence of beings. For Bachelard, 'poetics signal a relational dynamic between things', involving the vital dimensions of 'intimacy, secrecy, desire and repose' (1994, p. 222). The phenomenology of imagination is 'to capture images at their inception. … To imagine going down into water, wandering in the desert, is to change space – and to change space is to change being' (Bachelard 1994, pp. 221–223). Imaginations created through metaphor may give logos to perception, and more importantly nourishes an inner life (Kearny 2014). Groundwater narratives also build readers' understandings of groundwater's affects and agencies.

Bachelard (1994, pp. 187–188) cautions the poet, and I extrapolate to the climate writer, to beware the adjective 'ancestral', because as an adjective the word skips too quickly over the deeply imaginative places it refers to. 'Who knows the temporal dimension of the ancestral forest?' Bachelard (1994, p. 189) asks. Who really knows the temporal dimensions of groundwater? The metaphorical term 'ancestral waters', or 'dinosaur water', inferring complex groundwater flows over long timeframes, suggests imaginations of deep time geological, evolutionary and climatic events. Bachelard's ideas and expressions of the 'intimate spaces of immensity' infiltrate creative daydreaming about subterranean waters. I imagine an aquifer's immensity, its flows through deep time. Weaving a story from subterranean imaginings, finding metaphors to convey geologic depth, vastness and deep time, risks all the uncertainties of misfitting adjectives, slippery metaphors or aberrant similes. Climate writers face the paradoxes of intimate immensity at every writing session.

Difficult stories need to be told and retold. Depicting quotidian problems of groundwater extraction and pollution through groundwater narratives aims to avoid 'dystopia creep', a term which means a gradual acceptance of environmental conditions that were once unthinkable (Mills 2022, p. 39). A long-form poetic example of dystopia creep is Luke Best's narrative poem *Cadaver dog* (2020). Best articulates the horrors of fatal flooding, based on his experiences of the Toowoomba, Queensland 2016 floods. The poem describes the mission of the search and rescue dog whose task it is to find bodies. The narrator, a surviving mother, suffers grief-stricken paralysis, which relates not only to catastrophic flooding that has isolated and destroyed her home, but also to the unthinkable familial consequences of the flood (no spoiler). The muddying of personal grief and climate change induced deluge highlights the potential for dystopia creep to render readers as immobilised as the characters. In the following stanza from early in the narrative, Best makes a commentary on the power of story, the importance of retelling.

> And yet, events cannot evolve into story
> without their retelling, nor oats into gruel
> without
> water. So think of the mud as porridge, my yard
> as the bowl, the slope behind, just as it is, the house –
> the
> swollen Paddle-Pop shack it's become – as a
> dwelling-cum-
> deathtrap. Then think, why the cadaver dog. Why the
> story – its
> retelling.
>
> Best 2020, p. 8

Despite the value of retelling, dystopia creep hovers in the recesses of dark narratives.

Even as climate writing incorporates scientific knowledge, claiming accuracy in what may be labelled realistically framed narratives, it is not mimetic, and is not beholden to criteria of fidelity or a sense of singular truthfulness (Felski 2008; Mulhall 2009). This is discussed in more depth in Chapter 5. Rita Felski (noting Ricoeur's similar positioning) reminds writers that:

> we can question the politics built into certain storylines, as feminist critics have done ... , but our conceptions of politics, literature, human relations, the interaction between social structures and human agency, remain deeply beholden to the logic of narrative.
>
> Felski 2008, p. 85

It is by being 'beholden to the logic of narrative' that the materiality of narrative groundwater is linked with its storying capacity. Through storying *with* groundwater, telling an emotive story of its impacts on everyday lives, narrative groundwater remains beholden to wider historical, societal and environmental contexts for its logical expression. Storying *with* groundwater asks climate writers to challenge the prevalent political insouciance towards climate change catastrophes and groundwater's vulnerability.

Narrative groundwater has expressed itself in stories from many time periods. Groundwater spoke in the building of the Mesopotamian qanats, which are hand-dug underground tunnels that tap into deep wells in high grounds and move water by gravity often up to many kilometres to where the gradually sloping tunnel reaches the surface. Still under community management to ensure equitable distribution of water in Iran, the qanats are world heritage subterranean structures. Groundwater speaks through the sink wells of India. It tells stories around the wells in Saharan oases. Groundwater narratives emerge from any range of times and places, between historic hand-dug wells to contemporary highly engineered water-boring activities. Water engineering, well before it became a university qualification, has created sites for contentious groundwater narratives for centuries. Biblical stories of the well-building proficiency of Abraham and Jacob in the Old Testament, and the subsequent wealth it brought to their families, is an example mentioned briefly in a scene in *Why we cry*. Frankie, with a growing interest in groundwater, had joined the local choir and was talking with her friends after choir practice.

Excerpt 3.1: *Why we cry*, Chapter 4

Betty Winsome, the accompanist, wobbled towards them holding a teacup with a slice of cake teetering in the saucer.

'You three talking groundwater? Old as the hills.' Her face was flushed and sweat beads had formed under her eyes, threatening to roll down her pillow-like cheeks. Frankie, Wade and Lizard made room, could see she had something to say.

Betty spoke in a rush, keen to share her views and knowledge, not needing an invitation to begin. 'You can't tell me groundwater is a new thing. Way back, Abraham drilled the hidden sea. You just read Genesis.' Here she eyed Lizard, who was well known for his heathen ways. 'The Old Book tells you how Abraham dug wells. Limestone hillsides it says. Lined the wells with stone.' Betty's face beamed; her eyes shone. She was on a roll. Her chest rose like a bowl of pink blancmange with a deep fold down the centre, swelling above the swooping neckline of her cotton shift. She took a sip of tea, continued before her audience could do any more than nod.

'He lifted that old water for his family, his flocks, sheep and goats, from the deep. Made him very wealthy, influential, because of his access to water.'

'What's changed? Who hasn't heard of Abraham?' Lizard grated his hands together.

Betty shook her head and nodded almost simultaneously. She was determined to prove an important biblical point, but Lizard continued.

'Reckon he climbed down ladders on those damp white walls, or raised the dirt and rocks out on a levered pole with a wooden bucket on one end? How'd he lift the water?' Lizard asked. Betty was flummoxed by such technicalities.

'How'd he find it underground in the first place?' Frankie asked.

Betty continued, ignoring Lizard, addressing Frankie and Wade with her biblical facts. 'When Canaanites tried to take over his wells, Abraham negotiated with them to share the water. He gave them stock, a gesture of good will. Which of course confirmed his ownership of the wells, maintained his power, his herds, food for his family. He knew not to fight his neighbours.' Betty nodded as though it was all over. She ate her cake, looking for Frankie's response. They turned to Wade when he spoke.

'Underground water, it's as precious as the top stuff. Been found around here for tens of thousands of years. Even before your biblical stuff.' He shuffled, uneasy. Betty shook her shoulders, disgruntled, like a hen released from a sack.

'It's how much you take.' Lizard's voice was roughened like the sound of cicadas thick on summer air.

'Nobody shares their extraction rates from stock and domestic bores. No one knows who's taking what', Frankie stated. She knew she was avoiding Betty's point, but was not going to start discussing biblical incidents she knew nothing about. 'And there's more people now', she added.

'It's a pumping race now. Who can extract the most, the fastest, wins.' Lizard rubbed his nobbled hands together again.

'You don't fight your neighbours.' Betty's voice tightened, almost to a squawk.

'You don't shit in your nest.' Lizard scratched at his neck.

Wade nodded.

'Everyone's silence is at the cost of aquifers. Most bores are totally uncontrolled.' Frankie's voice wavered, her water glass slopping as she gestured with her arms. This was her first public articulation of her growing unease. Embryonic ideas, quietly drilled home by her father over decades, smouldered. Lennie had repeated his missive: 'Go to the source.' As a child, Frankie often repeated the phrase in her pre-sleep, listening to Lennie throw empty beer cans at the kitchen bin. Now she said it to herself on the conveyor line at work. Something of the tinder had caught a spark. Betty caught some of her verve and continued talking, excited, long-winded.

'To drink the deep waters of ancient aquifers is to taste life, life from before Jesus, before Moses.' Betty paused for breath.

'Before Indigenous peoples were here. Even prehistoric marsupials must have drunk it', Frankie said.

'Spot on', Wade added.

'You're talking like a prophet, Betty.' Lizard leant towards her, momentarily immobile.

Betty continued, not really heeding Lizard's comment. 'You know Isaac, son of Abraham, well, he re-dug his father's wells at Sitnar, Rehboth and then at Beer-Sheba. He made a covenant with God and the Philistines. Preserve groundwater. That's what he demanded. And for your information, a prophet isn't trying to foresee the future. They just remind you that if you continue on with your foolish ways, there'll be trouble. Do you know how many times the word well appears in the Bible?' She drew breath, challenging the group with a glare.

'No bloody idea.' Lizard shook his head. Wade and Frankie shrugged, looked at the floor.

'Preserve groundwater.' Lizard diverted the question, raised his glass in toast.

'We've got to think wider. You don't wreck your place.' Frankie was unused to the feeling of her words coming from her mouth unedited.

Wade continued to nod. 'Especially when a million other creatures share it, need it.'

'And the plants', Frankie said.

'Deep-rooted trees.' They were on a roll, their comments swirling like liquid in a glass.

'You'd think Kasha Duff would know better.' Lizard Green had to say what they all thought. Frankie shuddered, Wade rolled his eyes upwards, Betty stared into her cup.

'To groundwater, and all the aquifers.' Frankie lifted her glass and made to click glasses with Betty's teacup. Betty sipped her tea, grinned.

Later, in her kitchen, Frankie imagined her friends returned and in their homes, turning on taps to wash their faces before they went to bed.

Whatever timeframe or epoch an author chooses for storying groundwater's influences, groundwater narratives written now aim to illuminate groundwater's sustainability or the risks to sustainable supplies. Narrative groundwater speaks for the here and now, from the past, and/or addresses speculative futures.

I next discuss several examples of fiction and poetry discussing groundwater's storying capacities, particularly through its emergences in springs and wells. Marcel Pagnol's two-part novel *The water of the hills* (1962) gives groundwater a prominent place, showing its influence on human and non-human lives. The novel is set between the world wars in the south of France in limestone country, between the city of Marseille and the foothills of the Massif de L'Etoile. The novel shows the reliance of farming on springs emerging from the limestone shelf. It is, however, the subterranean flows of water, and their contrived absence, that drives the story. Structured as a two-part saga, 'Jean de Florette' and 'Manon des sources' ('Manon of the springs'), tales of deception and greed for groundwater abound. Jean de Florette returns with his young daughter Manon to a block of fertile land he had inherited from his mother, with plans to grow carnations and vegetables. Jean de Florette's neighbour Ugolin, the novel's antagonist, secretly blocks Jean's spring with concrete, with a view to forcing Jean away from the desired limestone land, so that Ugolin and his father can reap the benefits of the land. Ugolin harbours a secret love for Manon, hoping to marry her when she grows up. Groundwater's value and effects have front and centre stage as Ugolin's plan unfolds. Without the spring Jean works himself to the grave, carting water from other sources around the village. He dies devastated by his failure to produce food on his now arid plot. In the second section of the novel, set about ten years later, through Manon's discovery of the blocked spring, Ugolin's deceit is revealed. Ugolin takes his own life once he realises his actions are known and that Manon will never love him. Pagnol describes the limestone springs in terms of their agricultural and domestic uses for his human characters. The spring provides a scarce and precious water supply, with devastating impacts in its absence. This novel depicts groundwater's vital role in intergenerational and local community relations. Readers hear its tinkling and gushes through caves and springs, the deadly tones when the spring water is lost, and the liveliness of its return.

In contrast, two very different poets, W.H. Auden, writing in England in the mid-twentieth century, and Banjo Paterson, the Australian mid-to-late eighteenth-century poet, created expressions of groundwater, illustrating groundwater's potency in the poetic form. Auden's poem 'In praise of limestone' (1948) evokes the love of land in parallel with his adoration of a young boy's body, through the analogy of appreciation of underground water in English limestone country. The poem starts evocatively:

> If it form the one landscape that we, the inconstant ones,
> Are consistently homesick for, this is chiefly
> Because it dissolves in water. Mark these rounded slopes
> With their surface fragrance of thyme and, beneath,
> A secret system of caves and conduits; hear the springs
> That spurt out everywhere with a chuckle, ...

Auden's words interplay love and loss, yearning and respite, through pastoral-styled landscape descriptions. In a languid, almost conversational style the poem is centred in a present, a moment in time and space, where the surface and underground landscape evoke not only the porosity of limestone, and the watery spaces in limestone caves, but the human frailty of body and love. The poem draws allegorically from landscape to human foibles, needs and eventually love. It builds through metaphors of hills, valleys and limestone springs towards sensual appreciation of the young male body that Auden adores. The poem builds a world of human frailties and the inevitability of death. However it is the murmur of underground streams under limestone country that ties Auden's narrator to faultless love. The poem illustrates the power of allegory to create imaginings founded on groundwater.

> ... but when I try to imagine a faultless love
> Or the life to come, what I hear is the murmur
> Of underground streams, what I see is a limestone landscape.

In stark contrast in poetic form, with its rhyming couplets and iambic rhythms, Banjo Paterson's poem 'Song of the artesian water' (1896) provides another example of a poetic work that gives voice to groundwater. Paterson wrote the poem in the early years of the Federation drought (1895–1902). The poem responds to the turbulent colonial period and settler attempts to continue expansion of agriculture to drier regions. The poem adopts the perspective of white Australian pastoralists, showing their desperate attempts to find subterranean water to continue stocking remote landscapes with cattle and sheep. The poem is completely blind to Indigenous knowledges of subterranean waters. First Nations peoples had lived successfully for tens of thousands of years in the desert inlands of Australia finding waters beneath the surface long before colonialists dug their bores. Sites of cultural significance are not mentioned. Paterson recounts the story of groundwater's 'discovery' and extraction from deep, deep down below the surface by a steam-driven bore-drilling machine. The poem draws analogies to the devilish determination of bore drillers and pastoralists to find water, and to how perilously close to hell are the depths of the bore. Associations of the underground with death, burials and Satan are long held. Paterson's depiction of the Faustian pact of taking water from the devil is prescient. As contemporary pastoralists, with the support of governments, continue to attempt to control arid landscapes by using groundwater for irrigation and stock, the perils of excessive extraction are ignored (Powell 2012). In Paterson's poem access to artesian water is assumed to be a settler's right.

> Sinking down, deeper down,
> Oh, we're going deeper down:
> And it's time they heard us knocking on the roof of Satan's dwellin';
> But we'll get artesian water if we cave the roof of hell in.
> Oh! We'll get artesian water deeper down …

A Canadian well-borer, J.S. Loughead, had been employed in 1886 by Queensland grazier Simon Fraser to seek out artesian waters. The steam engine took weeks to penetrate 512 metres, to release 2.1 million litres of artesian water per day for thirsty cattle (Cathcart 2009, p. 173). Loughead was not 'the canny Scot' that Paterson refers to in his poem, but his legacy of drilling for and releasing artesian waters across marginal grazing land began what resulted in an ongoing agricultural reliance on subterranean water (Cathcart 2009). Paterson captures the settlers' desperation to find groundwater. When the water table is finally breached, under artesian pressure the liquid gold spurts fountain-like into the air. We hear groundwater's sibilance; we smell its emergence from deep, devilish places. The final two stanzas sing the jubilance of conquest, of finding artesian water:

> But it's hark! The whistle's blowing with a wild, exultant blast,
> And the boys are madly cheering, for they've struck the flow at last:
> And it's rushing up the tubing from four thousand feet below,
> Till it spouts above the casing in a million-gallon flow.
> And it's down, deeper down –
> Oh, it comes from deeper down:
> It is flowing, ever flowing, in a free, unstinted measure
> From the silent hidden places where the old earth hides her treasure –
> Where the old earth hides her treasures deeper down.
>
> And it's clear away the timber and it's let the water run,
> How it glimmers in the shadow, how it flashes in the sun!
> By the silent belts of timber, by the miles of blazing plain
> It is bringing hope and comfort to the thirsty land again.
> Flowing down, further down:

It is flowing further down
To the tortured thirsty cattle, bringing gladness in its going;
Through the droughty days of summer it is flowing, ever flowing –
It is flowing, ever flowing, further down.

The short-term glory of finding water to quench the thirst of cattle belies the longer-term damaging impacts of continued boring of the Great Artesian Basin by settlers. Paterson's prescience could not have imagined the ensuing century or more of drilling the Great Artesian Basin. Water sources across the Great Australian Basin have been pushed to dangerous limits more recently by deep drilling exploration and mining for coal-seam gas in significant agricultural and environmental areas (Powell 2012). Paterson's poem reminds climate writers of the fraught history of groundwater drilling and humans' reliance on the glimmering flows.

Dilemmas of Storying *with* Nonhuman Entities

Climate writers face dilemmas as they endeavour to invoke voices of a range of human and nonhuman characters that are not their own. Potential pitfalls and the hubris of appropriating the voice of groundwater need to be taken into account. Climate writers seeking to ascribe groundwater with its own liveliness might question their right to speak for a more-than-human entity. How might a human writer proceed, assuming to story *with* such otherwise 'silent' entities? Dilemmas and risks of the storying *with* processes in writing of the 'natural world' (which is almost impossible to define because of the pervasive infiltration of human detritus), include tendencies to centralise human-centred pursuits, and ignoring dispersive anthropogenic impacts and damages that predominantly reflect the dominance of Western, capitalist and/or colonial viewpoints and actions. The generic *anthropos* in terms such as anthropomorphism, anthropogenic and the Anthropocene obscures considerations of exactly which humans are causing the damage, and what lives and what ecologies and landscapes are bearing the consequences of specific human actions (Cohen *et al.* 2016).

Thinking of human and nonhuman lives in terms of planetary scaled occurrences (such as sea-level rise, ocean warming, global overpopulation, global pollution, a global water crisis etc.), takes writers beyond individually based, modernist views as a basis for narrating stories (Clarke 2015, p. 13). Imagining subterranean waters, waters moving through underground spaces, refuses in the main to be expressed as an embodied experience of individual humans. *Homo sapiens* did not exist in many of the geological timeframes that laid down the beds for current groundwater flows. There are difficulties in writing with groundwater's scale and temporality.

Writing with a consciousness of the scale of the Anthropocene also risks 'de-politicization' of the work (Clarke 2015, p. 13). Presenting globally scaled problems also runs the risk of alienating and disempowering writers and their reading communities as the dilemmas appear insurmountable (Hennessy *et al.* 2022). Groundwater narratives, in the context of the Anthropocene, remain tied to the human limits of thinking and writing, consciously working to politicise and specify responsibility for groundwater abuses. Despite the ubiquitous presence of the *anthropos*, groundwater narratives aim beyond the merely human. Narrative points of view of human and nonhuman characters contribute to polyphonous possibilities.

In facing ethical and writerly dilemmas, writers take responsibility for their endeavours, and for the political intent that drives the work. Writing self-conscious climate fiction in the Anthropocene cannot be 'idle chatter'; I locate myself as 'eco-prophetic witness' to the potency of loss of groundwater (Rigby 2009, 2015). Groundwater narratives aim to disclose the consequences of ecocidal human activities, and also to express our ongoing connectivity to groundwater's potency.

Climate writers may agonise over whether to write or not to write into and through these dilemmas. Immersing oneself in the catastrophes of the Anthropocene may be paralysing. We anticipate failures, know that our voices are partial and easily drowned out by populist media cacophonies. Climate writers engage in ongoing literary conversations within a longer-term creative trajectory, aiming to expand knowledge of nonhuman worlds through the stories we write (Carter 2015, p. 7). Collaborations between climate writers is an emerging option that fits well with Phillips and Bunda's (2018) first principle of storying – to nourish thought, body and soul. Hennessy and colleagues suggest that collaborative creative writing strategies have the potential to 'emotionally support and engage writers, primarily focusing on how narratives of hope and optimism might counter affective responses of anxiety, and the resultant solipsistic inertia or surrender' (Hennessy *et al.* 2022, p. 2). Trusting collaborations with human and nonhuman actants expands the imaginary field.

Writers choosing to involve nonhuman animals as agential actors in their stories engage with the bourgeoning field of animal ethics. Many scholars and writers have written into this field of enquiry. Franz Kafka's 'A report for an academy' (2007) stands as an iconic example of giving voice to an animal, an ape who talks of his apeness as he gradually adopts more human characteristics. John Coetzee's (2004) protagonist Elizabeth Costello likewise questions human and nonhuman animal relations when she questions mass animal slaughters for human consumption, and many aspects of 'the lives of animals' in her series of fictional lectures. Haley Singer's collection of essays *Abandon every hope: essays for the dead* (2022) likewise asks readers to witness and question industrial scaled killing of animals, to face and contemplate the suffering of animals. Karen Joy Fowler's novel *We are all completely beside ourselves* (2014), Danielle Celermajer's creative non-fiction *Summertime: reflections on a vanishing future* (2021) and Laura Jean McKay's novel *The animals in that country* (2020) also contribute to animal ethics conversations. Their writing complicates the standard view of human separation from and dominance over animal lives. They each demonstrate how being in the skin of nonhuman animals risks speaking *for* another species. They each invite readers to think *with* the animal characters. Ursula Le Guin takes this a step further in her short story, 'The author of the acacia seeds and other extracts from the Journal of the Association of Therolinguistics' (1984), exploring textual expressions for imagined ant language through acacia seeds. Writing the lives of nonhuman animals is not just creating cute or dangerous talking creatures; it entails acknowledging the sentience, the agency and the resistances of nonhuman animals to human-centred meanings.

Water divining is an aspect of groundwater extraction that consistently drew scepticism from each of the dozen or so hydrogeologists I spoke with during the research towards this book. Each expressed strongly that divining or water dowsing is predominantly hocus pocus, with little scientific verification of it as a method for detecting subterranean water flows. Water diviners or dowsers continue to claim their capacity to sense and locate subterranean water, and in some cases claim to know the quality of the yet untapped water, noting whether it is filled with salts or smells, or is potable, before it is accessed (Moore 2007). Water divining provides an example of multiple perspectives that can be considered in groundwater stories. It posed a dilemma for how I might convey this anathema to science. I chose to depict the mystery (or hoax) of water divining in *Why we cry* in an unresolved way. Frankie Pankhurst's father, Lennie Pankhurst, had practised water divining and part of Frankie's journey is to determine her own views on his revered accomplishments. Lizard Green, a paternal figure and friend, who had been friends with Lennie Pankhurst, wants to find out whether Frank has her father's 'gift'.

Excerpt 3.2: *Why we cry*, Chapter 7

Two days later Frankie met Lizard as planned on the corner behind the pub. As she climbed into his tinny Toyota ute, a familiar smell of old man and timber shavings assailed her. Lizard often made her feel nostalgic. ...

Lizard explained that he wanted to show her a favourite place, a place not grazed, nor razed, nor trashed by miners. 'Somehow they missed this spot. An unburnt forest is rare these days.'

A dry creek crept through open box-ironbark woodland. Even in the heat of summer there was a softness, a largesse in the sweep of the hillside. The limbs of ancient grey box trees reached for the pale blue sky. Herbs and grasses flourished, mid-story acacias bounced. A pair of long-billed corellas loafed in a hollow of an enormous dead bough, seeming to lean towards each other, wing feathers touching. Their pink-smudged faces and drooping beaks seemed melted by the simmering warmth. They flew off together as Lizard and Frankie approached.

An ephemeral creek bed was filled with smooth stones, rounded and scoured by a millennium of intermittent flooding flows. Caramel browns, siennas and dark greys, the rocks knocked against each other underfoot, sounding like lawn bowls. There had been water here, but for the past ten years, nothing. Frankie peered under stones hoping to find a frog, or insect worlds. Some were homes to centipedes and slaters.

'What are we looking for?' Frankie asked.

'Adam's ale. That's how I see it. Ageless and sweet.' Lizard was searching for ways to tell Frankie something, perhaps about his passion for groundwater. He rambled through a series of questions, half sentences, part formed ideas that together sketched a picture for Frankie of his ideas and thoughts. 'If Neptune is the god of under the sea, why isn't there a god for groundwater?'

'Maybe it's a godless place.' Frankie was still peering under stones.

'There's magic in the emergence of water from rock. Maybe there should be a god of aquifers. We just have to make it up.'

'Sure.' Frankie moved a larger rock with her toe.

'There's been droughts before, girlie. There'll be droughts again.' Lizard scratched at his scalp. Frankie quashed the retort that she was no longer a girl.

'Yes, but there are more people in the district now. Nearly as many as the sheep. And humans use more water.'

Lizard shot a glance at Frankie. 'It ain't just the town's bigger. All the hobby farmers around the edges, on their ten acres and that, they've all got alpacas or horses or donkeys, or think they can still water their gardens, have lawns like bowling greens.'

Frankie hesitated, thinking of the amount of water her horse, Carol, was drinking. 'And they're all getting Kasha to put down new stock and domestic bores', she added.

'It's drying out down there for a lot of reasons.' Lizard pointed to his feet.

'For stock and domestics, they don't need a licence, not for any quantity of water they take.' Frankie paused. 'But I think they have to register their bores with the water authority these days.' Frankie wanted to show Lizard what she was learning about groundwater. Cicadas buzzed.

'The rules are different everywhere, you know. Different regions, different states. Different countries', Frankie said.

'Adam's ale don't take any account of borders.' Lizard continued along the creek bed, scouting for signs of greenery. He looked up at the tree canopies, searching for new growth fed by deep-rooted tendrils that found deep water.

'That's right. Here you only need a licence for commercial use of bore water. The government monitors what's used for irrigation, factories, and industrial stuff. But not for private bores.' Frankie felt a small flush of confidence in her knowledge. Lizard raised his eyebrows.

'Yes, they monitor commercial bores. Government. Waterboys. When they have the time away from paperwork.' Lizard repeated Frankie's statement as if he'd always known it.

'That's what I just said, Lizard.' Frankie knew that Lizard repeated women's words, expected to be given credit that they were his. She continued. 'But the water authorities might soon start trying to monitor stock and domestic bores too.' They walked in silence along the dry creek bed.

'You know who controls the groundwater now, Lizard?' Frankie asked, but before Lizard could reply she added: 'Aquifers cross borders. Surface borders mean nothing to an aquifer. It's not simple.' She drew a line in the air, so they both knew she was talking about the arbitrary lines on maps.

'Never was.' When Lizard sat down on a log, scratched his head, Frankie knew he was going to take the long way round in this conversation. 'Betty isn't the only one to know that biblical stuff.' He paused, plucked grass stalk, sucked the end. 'Isaac moved his clan to the Well of Lahairoi. Translation, Well of the Vision of Life.'

'Betty told you that.' Frankie swiped at a fly.

'Water was their life stream.' Lizard looked at the sky.

'Still is.' Frankie wandered towards a large grey box, the trunk wider than her arm span. She peered up into its spreading boughs, dizzied by the swaying limbs.

Lizard continued. 'We don't see it that way nowadays. Take it for granted. Mostly people don't even know where their water comes from.'

They walked on in silence, blow flies cutting the air, heat trickling into their bones.

'Aquifers get whatever's left over, after everything else has a go', Frankie mused.

'This is not the River Jordan. Imagine the rows over that water.' Lizard sat down again on a fallen branch. 'Palestine and Israel draw on the same aquifers that feed the rivers that flow through both their territories.' He paused while Frankie turned towards him, impressed. 'Israel diverted a whole river out of Lebanon for themselves. Jordan and Saudi Arabia both pump from the Disi aquifer for agriculture and for their cities.' Lizard took a deep breath.

'Makes our state water arguments look like playground spats', Frankie said, watching Lizard searching the inside of his skull for what he was trying to express. She continued before he could find it. 'Where the wells and bore sites are found is based on a combination of geology, economics, politics, laws, right?'

Lizard nodded.

'So how did those Bible people work all that out? Before the maps. It's bureaucrats and accountants in offices controlling things nowadays.'

'Where's the moral considerations?' Lizard ignored Frankie's question. 'Some people bloody pump aquifers to top up their swimming pools.'

'And they put up a sign to crow about using groundwater on their green, green gardens. As if it's theirs to take', Frankie fumed.

'Nobody thinks of it as a part of the commons.' Lizard sounded uncannily like her father speaking.

'The commons. For the common good.' Frankie repeated what she had so often heard Lennie chant.

'Why don't we remember the natural stuff, ecosystems and the like, that rely first and foremost on groundwater?' Lizard gulped his growing agitation, tightened his lips ready to sprout his outrage.

'Lizard, what? What's the matter?' Frankie stood near.

'We've just lost the ways of finding ...' Lizard paused.

'What are you trying to say, Lizard?'

'You know your dad was a diviner?'

Frankie nodded.

'Well, I always wondered if you had the gift too.' Lizard looked sheepishly from under his hat. 'I've got some sticks for you to give it a go.'

'Lizard! Hell. I don't know.'

'Go on Frankie, give it a shot. Along a dry creek bed is as good a place as any to get a tweak. They say the water flows under the surface in the ancient stream beds. Deep leads. Electro-magnetic fields give them away. Reckon you can find it? You got the touch.' Lizard half grinned, enticing, provocative, nervous at once.

'Hell, Lizard. I thought you were going to tell me something I didn't know about Dad. Some long-lost secret.'

'Maybe it is something you never knew. We'll find out if he gave you the touch. The touch for water.' Lizard held out two sticks with the bark rubbed off, smoothed to a low sheen. Their surface patina attracted Frankie and she reached to touch the wood. Lizard handed them over. Frankie examined the sticks, fingering the tactile curve and flow of the wood grain.

'Where'd you get these sticks?'

'Found 'em. At Lennie's place. He said I could borrow them.' Lizard sat stock still.

Frankie knew that Lizard had occasionally visited Lennie at the cabin, to fish and chew stories with his old mate.

'These are Dad's dowsers?'

'Thought you'd better try with the best.'

Frankie smoothed the sticks in her hands, recalling her father's nobbled knuckles rubbing them down with an oily rag. She saw an image of

Lennie threading his way across a paddock, led by his sticks out front, head down concentrating on the deep cycles of earth, its magnetism, its deep time stories. She'd heard that Kasha kept a pair of wires under the front seat of his bore rig truck. Brought them out each time he pulled up for a dig. She was convinced he was making it all up. She stared at her father's divining sticks, remembered playing with them as a child, Lennie shouting at her to put them down. 'They're not toys, you know', he'd bellowed.

'OK.' She sighed deeply. 'But don't expect anything. It's probably bullshit.' She remembered her geology teacher scoffing at the suggestion that diviners could really find water underground. He told the class that diviners worked a few short tricks, looked at the landscape for the most likely place to wave their wands. Some even rang up hydrogeologists to get the overall locations of water before they went on to a site to hoodwink the landowners. Lennie had always scanned the landscape when he went dowsing, went to the lower spots in the land, looked out for the biggest trees that might be tapping into aquifers. But what made his sticks dance in his fingers like puppets? A memory of his twitching hands gave Frankie a tingle up her back and neck.

Frankie held out the sticks, balancing them between her thumb and the middle joint of her index finger, as she'd seen Lennie hold them.

'You've got it, love. Go on.' Lizard encouraged her with a gesture towards the creek side.

Frankie walked slowly, balancing the baton-like sticks in her hands. Her skin touched them lightly. She wondered what she might feel. She placed each foot cautiously, stepping over rocks and logs, sticks crackling, long grasses waved seed heads and bumped her shins. Birds called her approach. She froze when a brown snake slithered away, disappearing into a crevice. She turned in the opposite direction still holding the sticks out front, till her arms ached and beads of sweat dripped down her back.

Lizard was perched on a fallen log in the shade of a tree on the edge of the creek bed, watching her every movement from under his hat.

'This is hopeless. I don't know what I'm doing', she called, stretching her arms.

'Keep at it, love. It's an ancient art.'

'Come on, Lizard. Let's go home. These are Dad's sticks. It's not my thing.'

As Frankie approached Lizard, she held out the sticks one last time. The air thrummed with cicadas, dry heat vibrated. Perhaps her toe stubbed a rock, perhaps she stood on a twig which cracked and made her jump. The sticks twitched and wavered, pulled her hands towards the ground. She let them drop.

'You've found it. You've got it.' Lizard leapt to her side.

'I just tripped, Lizard.' Frankie collected and clutched the sticks together.

'Come on, dig here.' Lizard wouldn't take the sticks, was on his knees, scratching at the ground, pulling at a large stone. As he heaved it from the earth's hold a dribble of water pushed into the hole. 'I knew you could do it. I knew.' Lizard stood up beaming and held Frankie's shoulders with his dirt-covered hands. 'Look, you found it!' Lizard jigged a little and bent to cup his hand into the tiny puddle.

'It was always there. We're in a dry creek bed. Look, the hills around mean that this is where the water has to collect.' Frankie waved the sticks in front of her face to hit away the flies as well as dismiss Lizard's claim to her success.

Lizard let the dribble of muddy water run through his fingers. 'Wish I could tell that father of yours.'

Frankie flushed, strutted away, heading back towards Lizard's car. 'Come on, I've got uni work to finish. A geology assignment, actually. This is crap.'

Lizard rubbed the sludge between his hands, wiped it onto a rock. He followed Frankie, muttering, 'I told that old bastard that she had it. I told him.'

Depictions in *Why we cry* of water divining and the mysterious processes of finding subterranean water through feelings in the hands, sticks or wires were influenced by my reading of Randolf Stow's *Tourmaline* (Stow 1963/2015). Stow's protagonist, Michael Random, is said to be a water diviner, and is ascribed mystical qualities by the residents of the dusty, drought-stricken mining township Tourmaline. He finds gold instead of water. Random takes on Christ-like qualities as he leads spiritual rituals in the remnants of a falling-down church. Ideas of belief and spirituality, the inner workings of humans and the subterranean are implied by Stow in his depictions of Tourmaline's residents' various attitudes to water divining.

First published in 1963, Stow's novel places a raft of diverse characters in a drought-ridden setting. The final scene depicts a massive dust storm. The narrator disappears 'into the thick red wind' and nearly drowns in the dust as the bell in the disused church tolls (Stow 1963/2015, p. 251). Stow demonstrates that 'matter', whether it be dust or gold or water, can permeate a narrative, get under the skin of characters and readers alike. Though not touted as climate fiction, Stow's novel pivots around a sense of catastrophe in divine intervention that masks human turpitude. Stow gives no scientific rationale for the drought nor Michael Random's inability to find groundwater. Nevertheless, *Tourmaline* remains one of the few Australian novels to consider the place that groundwater might have in the Australian psyche.

Another Australian novel that subtly shows the presence of groundwater is Elizabeth Jolley's *The well* (1986). Likewise, this novel does not generally qualify as climate fiction. It has merely two sentences linking groundwater to the well around which the story hovers.

> In the distance she saw the line of trees which, her father always said, must thrive on an underground water supply, and which marked the furthest end of what used to be her property and where the dog-leg was. Seeing the trees even though they were a long way off reassured her.
>
> Jolley 1986, p. 146

The well in the backyard of the rural homestead is where Jolley's protagonists, Hester and Katherine, dump the body of the ambiguous creature/human they had knocked with their car. From the well the young and vulnerable Katherine hears voices of the supposedly dead creature/man. Groundwater is presented indirectly through its evidence in the depths of the well. It is both a source of life and death. The well, which would probably have been connected to the previously mentioned groundwater, carries overtones of fear of death. Jolley does not articulate the connection between the reassuring tree line, nourished by groundwater, and the well in the house's yard. This common dissociation of springs and wells from their source, and from their effects where they emerge at the surface, is the disjuncture that groundwater narratives explore.

As part of the research for this book I undertook a field trip to the mound springs of South Australia as a way to experience first-hand the emergence of ancient groundwater culminating its 2000-year journey from Queensland towards South Australia through the belly-like depths of the Great Artesian Basin. These experiences led to passing references to mound springs in *Why we cry*, and to a deepening sense of the timeliness of groundwater narratives. The following excerpts from a memoir-styled essay provide an example of storying with groundwater through creative non-fiction. Science enables the 'behaviour' of groundwater to be braided with the author's experiences of seeing 'dinosaur water', ancient groundwater emerging at the mound springs of South Australia. This is a personal story, a short memoir of a month-long journey to the springs of the Oodnadatta Track and to Dalhousie Springs in South Australia.[6] As an example of creative non-fiction, many of the elements of storying and climate writing are practised in this piece.

Excerpt 3.3: 'Groundwater'

The water is the temperature of blood. Under a white-hot sky the swimming hole at Dalhousie Springs is fringed by thick reeds and creamy-flowered, paper-barked melaleucas. Long-billed corellas on drooping branches watch me float. Slightly saline, a whiff of sulphur suggests the decades, centuries, perhaps the millennia that this water has been seeping through aquifers of the Great Artesian Basin. The relief of bathing in silky warmth lulls me to thinking that there's plenty of water after all. After days of dusty driving, I am thinking about the sustainability of groundwater. What is my relationship to this warm, watery matter? I tell the story of this journey to the mound springs as a way to invoke a connection, a relationship to the hard-to-see places. This journey is taking me to places where I understand myself more deeply. Seeing the trickling emergences of water in the desert reminds me of the importance of an inner life, something beyond the material, below surfaces. ...

We camp about five kilometres away from the bustling campground, choosing instead the sandy bed at Three O'Clock Creek. Red mulga, a hardy acacia with distinctive maroon curls for bark, grows along the creek bed. Known locally as miniritchie, red mulga's strong limbs provided valuable hardwood for tools and weapons, later fence posts and stockyards. Its companion species in the dry creek bed, the gidgee and coolabah trees, provide much needed shade, as the land around is bare and pulsing hot. I listen for the *maw-ww* of the waterbag frogs, who burrow in the river sands, waiting in semi-hibernation at times for years. They crawl out for breeding when the water flows. The twitter of honeyeaters and the rasp of crickets and cicadas rob the shimmering heat of its silence. I wonder about the water that dropped far beneath the gravelly sands of the dry creek bed. How deep did the waterbag frogs bury themselves with their cache of water stored in pockets of their skin? Dry spells are long. The mulga, gidgee and coolabah roots channel water from way beneath the waterbag frogs. Water cycles continue.

We return to Dalhousie Springs each afternoon to swim in the earth-warmed pool. Each day about four megalitres of tepid water rises naturally under artesian pressure from nearly two thousand feet below to fill and overflow from Dalhousie Springs. That's four Olympic-sized swimming pools of groundwater flowing upwards through hair-like pores in the crust. The entrance to the main pool sports signage by Indigenous park managers welcoming visitors. Again I feel like an intruder. A splash of colourful polystyrene noodles and a bright-orange life-saving ring decorate the decking. We climb down the wooden steps to flop into the reed-lined pool, relieved to wash away campfire smoke, sweat and dust. I float, torpid, motionless, watching with one eye the long-billed corellas who perch like sentinels on high. Unique to these springs, a tiny fish named the Dalhousie guppy nibbles at my skin, keeps me awake. In water at thirty-seven degrees I soon melt, soporific. Ancient waters seep into my skin, bringing slow meanings to presence, to connections to place. I climb from the pool in only my undies as other travellers arrive for a dip. The warm water leaves me languid, calm, strengthened. As the earth absorbs waters through layers of skin-like sediments, I carry her waters in my skin. Waterbag me.

Weaving over sand tracks through the complex scattering of springs we wander back to camp. A dingo in the distance pounces on prey in the long grass, trots away head high, successful. This feels like sacred space. I understand something more of the land. ...

We re-trace our route southward, remembering, re-visiting campsites. Familiarity with the track reassures me. I try to imagine knowing this route as intimately as generations of Indigenous travellers before me knew these ways to water. Their wisdom grows from telling and re-telling stories of the land. Profound stories emerge from their attentive knowledge. The cycle is firm. Deep accountability to water and country is enmeshed. Had we passed through, driven onwards, and crossed the Simpson Desert with a wobbling orange flag on a pole, to return by some other route, I would perhaps have missed this echo-like perspective that resounds, enabling repeated experiences to permeate. The return journey adds depth to what I'd found, lessens the transitory feel to the journey. ...

Swimming in warm spring water under open desert skies focuses my respect for this country. I drink spring water into my cells, imbibing a legacy of aquifers across millennia. Travelling to the mound springs kindles my responsibilities to preserve groundwater. The waters that hold me afloat inspire me to listen to the potency of subterranean stories. Voices of aquifers had called me on this odyssey, a journey to little-known places. Rather than as a resource to be extracted whatever the cost, I learn of groundwater's place in the inherent psyche of this continent. Aquifers beneath our feet remind me of a life force beneath my skin. The places that tingle when a thing of beauty shines.

In witnessing the wavering pulse of the mound springs, I lament the lost potency of subterranean realms. I am enraged at the lack of political courage to address hard questions. Failures to consider the long-term effects of extracting and polluting groundwater for short-term profits are incomprehensible, inexcusable. Below the surface, ancient aquifers are the lifeblood of futures. One way to build their sustainability is to tell their stories. The unique ecologies of mound springs are gasping. My attention has been turned to deep places. Mound springs are on life support.

The processes of storying groundwater's many iterations can take innumerable forms. Between the examples of fiction, creative non-fiction and poetry mentioned in this chapter lie many other options. What remains significant is the author's considerations of the principles of storying as a foundation to thinking into various storying forms. Storying *with* emphasises the enmeshments we have with the more-than-human world. It nourishes thought, body and soul, and claims new voices in the silenced margins (Phillips and Bunda 2018, p. 45). Groundwater narratives emerge from practices of storying *with* to expand readers' connections to subterranean waters. As will be discussed in Chapters 6 and 7, it is how humans come to be affected by worlding of groundwater's presence through stories that gives rise to agential politics, to capacities for meaningful actions. The next chapter examines in more detail the ways that groundwater narratives interact with the science of hydrogeology.

Notes

1 Tracey Bunda and colleagues created an exhibition titled *Stories of Belonging* with Indigenous and non-Indigenous artists/scholars for an arts festival in Queensland, Australia. Storying processes underpinned the research methodology of the collaborative exhibition and the ensuing publication (Bunda *et al.* 2019).

2 Country is a specific word to incorporate connection to ancestral lands in Australia, connection to the rocks and trees, rivers and mountains of that land. Country vitally includes the cultural practices and stories of a place (Wright 2016).

3 Other award-winning and outstanding Australian Indigenous writers, poets and musicians to consider include William Barak, Melissa Lukashenko, Archie Roach, William Barton, Natalie Harkin, Samuel Wagan-Watson, Evelyn Araluen, Lou Bennett, Alexis Wright, Kim Scott, Larissa Behrendt, Alison Whitaker and Marcia Langton, to mention only a few who have and continue to speak out for Indigenous justice.

4 Amanda Boulter's *Writing fiction: creative and critical approaches* (2007) is one of many useful examples.

5 The complex philosophical and linguistic discussions that followed Lakoff and Johnson's (1980) book *Metaphors we live by* are beyond the scope of this book. Haser (2005) provides an excellent overview and critique of Lakoff and Johnson's oft quoted discussions of conventional, ontological and conceptual metaphors, from a deconstructivist, philosophical line of argument.

6 The full version of the essay first appeared in *Meanjin* Vol 82. No. 2(2) (2023). *Meanjin* kindly granted permission to include these excerpts.

References

Admussen, N., 2016. Six proposals for the reform of literature in the age of climate change. *The Critical Flame: A Journal of Literature and Culture*, 42, 1–6.

Auden, W.H., 1948. In praise of limestone. *PoemHunter* [online]. Available from: www.poemhunter.com/poem/in-praise-of-limestone-3/ [accessed 16 August 2022].

Bachelard, G., 1994. *The poetics of space*. Boston, MA: Beacon Press.

Bakhtin, M.M., 1981. *The dialogic imagination: four essays*. Austin, TX: University of Texas Press.

Bakker, K., 2010. *Privatizing water: governance failure and the world's urban water crisis*. Ithaca, NY: Cornell University Press.

Barnes, T.J. and Duncan, J., 1992. Introduction: writing worlds. *In:* T.J. Barnes and J. Duncan, eds. *Writing worlds: discourse, text and metaphor in the representation of landscape*. London: Routledge, 1–17.

Baucom, I. and Omelsky, M., 2017. Knowledge in the age of climate change. *South Atlantic Quarterly*, 116, 1–18.

Best, J., 2020. *Cadaver dog*. Brisbane: University of Queensland Press.

Birch, T., 2016. Climate change, mining and traditional Indigenous knowledge in Australia. *Social Inclusion*, 4, 92–101.

Birch, T., 2020. Recovering a narrative of place. *Griffith Review – First Things First*, 60, 207–219.

Boulter, A., 2007. *Writing fiction: creative and critical approaches*. Basingstoke: Palgrave Macmillan.

Bunda, T., *et al.*, 2019. Storymaking belonging. *Art Research International: A Transdisciplinary Journal*, 4 (1), 153–179.

Carter, P., 2015. *Places made after their stories*. Crawley, WA: University of Western Australia.

Cathcart, M., 2009. *The water dreamers*. Melbourne: Text Publishing Company.

Celermajer, D., 2021. *Summertime: reflections on a vanishing future*. North Sydney, NSW: Hamish Hamilton.

Chen, C., Macleod, J. and Neimanis, A. 2013. Introduction: towards a hydrological turn? *In:* C. Chen, J. Macleod and A. Neimanis, eds. *Thinking with water*. Montreal, QC: McGill-Queen's University Press, 3–22.

Clarke, T., 2015. *Ecocriticism on the edge: the Anthropocene as a threshold concept*. London: Bloomsbury.

Coetzee, J.M., 2004. *Elizabeth Costello*. London: Vintage.

Cohen, T., Colebrook, C. and Miller, J.H., 2016. *Twilight of the Anthropocene idols*. London: Open Humanities Press.

Danielewski, M., 2014. Foreword. *In:* G. Bachelard, *The poetics of space*. New York: Penguin Classics.

Davis, M., 2020. The long road to Uluru. *Griffith Review – First Things First*, 60, 13–45.

Felski, R., 2008. *The uses of literature.* Malden, MA: Blackwell Publishing.

Fowler, K.J., 2014. *We are all completely beside ourselves.* London: Serpent's Tail.

Haser, V., 2005. *Metaphor, metonymy, and experientialist philosophy: challenging cognitive semantics.* Berlin: Mouton de Gruyter.

Hennessy, R., Cothren, A. and Matthews, A., 2022. Creating new climate stories: posthuman collaborative hope and optimism. *TEXT*, 26, 1–20.

Janke, T., 2019. *Protocols for using First Nations cultural and intellectual property in the arts.* Pyrmont, NSW: Australia Council for the Arts.

Johnson, A., 2021. Writing ecological disfigurement: First Nations poetry after 'the black grass of bitumen'. *TEXT*, 25, 1–25.

Jolley, E., 1986. *The well.* Camberwell, Vic: Penguin.

Kafka, F., 2007. *Metamorphosis and other stories.* London: Penguin.

Kearny, R., 2014. Introduction. *In:* G. Bachelard, *The poetics of space.* New York: Penguin Classics.

Lakoff, G. and Johnson, M., 1980. *Metaphors we live by.* Chicago, IL: University of Chicago Press.

Le Guin, U., 1984. The author of the acacia seeds and other extracts from the Journal of the Association of Therolinguistics. *In:* U.K. Le Guin, *The compass rose.* London: Panther Books, 11–19.

Leane, J., 2016. Other peoples' stories: When is writing cultural appropriation? *Overland*, 225, 41–45.

Ma, N., dir., 2015. *Putuparri and the rainmakers.* Australia: Ronin Films.

MacFarlane, E., 2013. *Reading Coetzee.* Amsterdam: Rodopi.

Mahood, K., 2022 *Wandering with intent: essays.* Brunswick: Scribe.

Marshall, V., 2017. *Overturning aqua nullius: securing Aboriginal water rights.* Canberra: Aboriginal Studies Press.

McGraw, J. and Pieris, A., 2015. *Assembling the centre: architecture for Indigenous cultures. Australia and beyond.* London: Routledge.

McKay, L.J., 2020. *The animals in that country.* Brunswick, Vic: Scribe.

Mills, J., 2022. Dystopia creep. *Meanjin*, 80, 38–46.

Moggridge, B., 2010. *Aboriginal knowledge and cultural values of water.* Canberra: CSIRO.

Moore, A., 2007. *The wisdom of water.* Castlemaine, Vic: Python Press.

Mulhall, S., 2009. *The wounded animal: J.M. Coetzee and the difficulty of reality in literature and philosophy.* Princeton, NJ: Princeton University Press.

Pagnol, M. 1962. *The water of the hills.* London: Pan Books.

Pascoe, B., 2014. *Dark emu.* Broome WA: Magabala Books.

Pascoe, B., 2020. The imperial mind. *Griffith Review – First Things First*, 60, 234–243.

Paterson, B., 1896. Song of the artesian water. *Poetry Library* [online]. Available from: www.poetrylibrary.edu.au/poets/paterson-a-b-banjo/song-of-the-artesian-water-0004015 [accessed 16 August 2022].

Phillips, L.G. and Bunda, T., 2018. *Research through, with and as storying.* Abingdon: Routledge.

Powell, O.C., 2012. Song of the artesian water: aridity, drought and disputation along Queensland's pastoral frontier in Australia. *Rangeland Journal*, 34, 305–317.

Rigby, K., 2009. Writing in the Anthropocene: idle chatter or ecoprophetic witness. *Australian Humanities Review*, 47, 1–10.

Rigby, K., 2015. *Dancing with disaster: environmental histories, narratives and ethics for perilous times.* Charlottesville, VA: University of Virginia Press.

Rose, D.B., 2013. Slowly – writing into the Anthropocene. *TEXT*, Special Issue 20.

Singer, H., 2022. *Abandon every hope: essays for the dead.* Perth: Upswell.

Sontag, S., 2007. At the same time: the novelist and moral reasoning. *In:* S. Sontag, *At the same time: essays and speeches*, ed. P. Dilonardo and A. Jump. New York: Farrar, Straus and Giroux, 210–230.

Stow, R., 2015. *Tourmaline.* Melbourne: Text. (First published 1963.)

Todd, Z., 2015. Indigenizing the Anthropocene. *In:* H. Davis and E. Turpin, eds. *Art in the Anthropocene: encounters among aesthetics, politics, environments and epistemologies.* London: Open Humanities Press, 241–254.

Tynan, A., 2020. 'Que peut la fiction?' Storying the unexperienced experience in Jorge Semprun's fiction. *Modern Language Review*, 115, 46–62.

Van Neerven, E., 2019. Restorying care. *Overland*, 235, 12–17.

Wright, A., 2016. What happens when you tell somebody else's story? A history of Aboriginal disempowerment. *Meanjin*, 75, 58–76.

4

HYDROGEOLOGY

GROUNDWATER NARRATIVES AND HYDROGEOLOGY

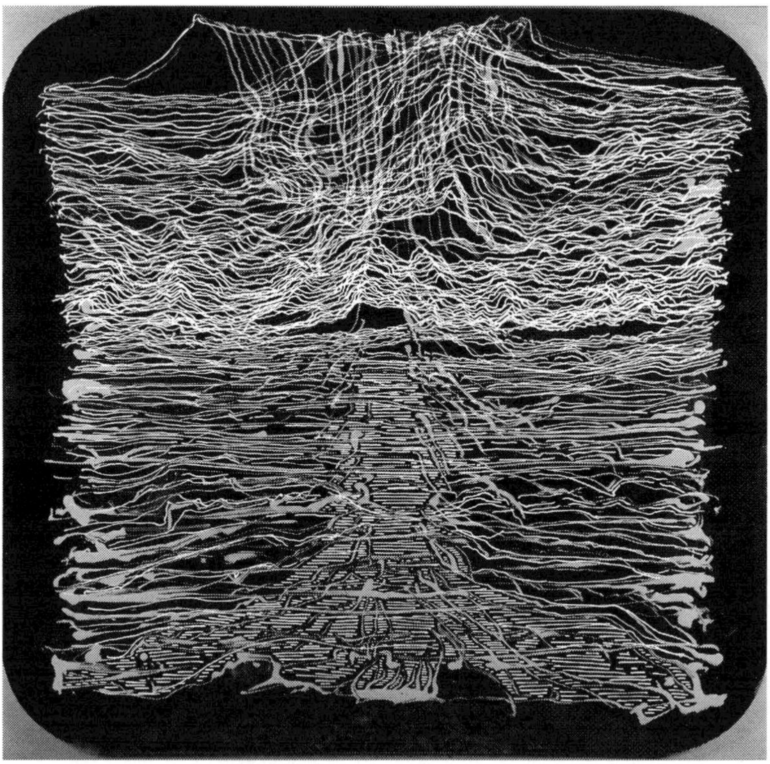

Figure 4.1 Bill Sampson, *Drawing the line*, 2022, synthetic polymer and enamel on aluminium, 40 × 40 cm.

DOI: 10.4324/9781003270416-4

Hydrogeology is the systematic study of groundwater, including its various movements through and relationships with rock. Hydrogeologists develop models to calculate/estimate predictions of groundwater's volumes, movements and constituent chemistry. Hydrogeologists' foci range from the microscopic pore spaces in various porous rocks to wider landscape-scale viewpoints as they analyse groundwater's multiple characteristics. Hydrogeologists work with many other hydrological specialities, including hydrometeorology, climatology, hydrography, geomorphology, limnology and oceanography.

> [M]ay there not come that even bolder adventurer – the first geolinguist, who, ignoring the delicate, transient lyrics of the lichen, will read beneath it the still less communicative, still more passive, wholly atemporal, cold, volcanic poetry of the rocks: each one a word spoken, how long ago, by the earth itself, in the immense solitude, the immenser community, of space.
>
> Le Guin 1984, p. 19

Foundational to groundwater narratives is a writer's engagement with the science of groundwater, hydrogeology. Groundwater scientists develop knowledge about groundwater through scientific research methods. They share understandings of groundwater issues with their community of scientists, with employers, with groundwater management and policy regulators and with wider community stakeholders. Groundwater may be seen as a 'quasi-invisible resource', making the communication of its features a specific challenge (Richard-Ferroudji and Lassaube 2020). Reading hydrogeological texts and journal articles, alongside pursuing conversations with practising hydrogeologists, is one way to start to appreciate the complexity of the science specialities that address groundwater's issues and contexts. We become immersed in a new language. In 'The Author of the Acacia Seeds', Ursula K Le Guin (1984, p. 11) invites readers to inhabit the 'dialect of Ants'. She builds a story of ant communications and finishes the imaginaries of therolinguistics with the invitation to geolinguists to read beneath the surface languages. In the interplay between hydrogeology and climate writing, the shared ground – the places where thinking, scientific knowledges and writing processes meet – becomes what Donna Haraway calls the dilly bag, the holy grail, the carrier bag for situated knowledges (Haraway 1988).[1] In groundwater narratives we become hydro-geolinguists, expressing the languages of science and subterranean waters.

Hydrogeologists may specialise in a range of scientific disciplines, from water chemistry, physics and hydraulics of water's movements, through computer modelling and imaging, to the ecologies and biology of groundwater-dependent ecosystems. Hydrogeologists primarily study geology – the formations of rock types, crystal and grain chemistries of various rocks and their porosity and compressibility in various conditions. They must also understand water – its atomic and chemical behaviours, its compressibility, its responses to moving through various rock types over large or small distances, its viscosity at various temperatures and chemistries, its responses to gravity and hydraulic head. Why does an artesian bore spurt fountains of water twenty, forty, sixty metres into the air? Hydrogeology has a strong mathematical element, calculating water's flow rates through different porosities and permeabilities of rock, though confined and unconfined aquifers. Models estimate flow rates in multiple dimensions. Significantly, groundwater's relationships with surface waters is an emerging body of knowledge. Deep underground, groundwater might plume vertically (up and downwards) as well as horizontally. Water responds to different geological formations as it moves, usually infinitesimally slowly towards oceans, or whatever discharge points or regions they first reach.

Hydrogeology, along with the complementary fields of groundwater law and policies, and water engineering, which designs methods for groundwater extraction, provides a foothold for writing, reading and analysing groundwater narratives. Thinking and research within cultural and creative scholarship in the environmental humanities is expanding ways to work collaboratively with science knowledge. This work aims to develop wider community responses to climate change. The research and writing in various forms of creative climate writing bring together the edges of fact and fiction. Phillip Gerard, in his discussion of creative non-fiction, argues that writing with a solid grounding in factual events or circumstances is vital, but ultimately is only as interesting as the writer/poet makes it in terms of larger and more meaningful concepts that 'resonate beyond the historical moment' (2018, p. 3). Contemporary climate writing, which responds to the specific demands of anthropogenic damages to the planet, is charged not only with conveying or disseminating information about the litany of catastrophes occurring, but also with giving readers an avenue into understanding both human culpability and potential ways forward.

The complexity of climate change is shown in its many scientific specialisations, sometimes known as knowledge silos, including meteorology, climatology, hydrogeology, oceanography, geology, atmospheric chemistry and glaciology. Each field of science contributes understandings to the mosaic-like picture of causes and effects of global warming. Hydrology, meaning the study of water, also has many sub-disciplines that lead to specific knowledge bases – these include hydrogeology, which puts geology and water in tandem. Hydrometeorology, limnology, hydrography, geomorphology, hydrogeochemistry, hydroecology, hydraulic engineering and hydro-informatics are all specialities in this space.

Michael Segal (2017) argues that affective climate fiction can provide bridges between the 'laboratories' of science and culture and reveal this to the wider public. Segal (2017, p. 127) suggests that inviting science into messy conversations in the realms of culture and politics is a necessary step in this bridge making. Predictions that groundwater will play an essential role in human adaptations to climate change and the impending world-wide water crisis are an important motivator to this book (Brodie *et al.* 2012; Rockstrom *et al.* 2023). Writing *with*, or storying evocatively around the ideas of groundwater's subterranean presence, is here founded on the notion that groundwater's presence and movements are only partially predictable through the current limits of sciences. Mathematical representations and modelling of groundwater belie its incalculability.

Scientific narratives, with all their imperfections and opportunities to grow understandings of the world, have influenced the ways humans understand themselves (Clarke and Rossini 2011). We know ourselves to be insignificant, ant-like in the swirling galaxies that astronomy has revealed. Through biological and medical sciences, we know ourselves as genetic conglomerations with long inherited lineages. We have become cellular, chemical bodies that can be treated with pharmaceuticals. As bodies of water, we are aqueous, connected to the hydrosocial cycle and knowable through what we can know of water.

In an era of recognized climate change, science narratives enable an increase in understanding of distinctions between anthropogenic causes and effects and epochal climate impacts. Science narratives of many types give a basis to making meaning. As Michael Segal says:

> Scientific narratives are some of the most powerful of all. They teach us more than facts, mechanisms, and procedures. They convey a world-view of skeptical empiricism and indefinite revision, show us how to negotiate the boundary between our rational and emotional selves, teach us to suspend judgment and consider all the possibilities, and remind us that a belief in objective truth is a deep kind of optimism with massive dividends. Perhaps most important of all, they situate us in the world.
>
> Segal 2017, p. 123

Traditional Western science narratives run the risk of being part of the colonialist project, supporting the view of the world as only a resource to be exploited. Indigenous sciences are slowly being recognized in many Western science fields. Indigenous knowledge fields include astronomy, ecology, botany, agriculture and medicine (Nakata 2002; Hamacher and Anderson 2022). It is a trap to think that the scientific revolution is the sole domain of white men. Indigenous cultures worldwide give vitality and meaning to their landscapes through long-held research and knowledge transmitted across generations though daily life practices, and through stories, songs and dances. Recognition that landscapes are neither inert nor mute, but capable of expression of their vitality, has underpinned not only rich cultural expressions of landscape, but has ensured care and responsibility for forests, waterways and lands over time. Amitav Ghosh makes this case in detail with reference to Dutch colonising of the East Indies, specifically the Spice Islands of Maluku, in his creative non fiction book, *The nutmeg's curse: parables for a planet in crisis* (2021). Colonisers saw nutmeg only as a valuable commodity to take to Europe, with little recompense to the islanders. Giving it the name *Myristica fragrans* Houtt silences the centuries of local names and stories that this fragrant fruit carries for traditional owners of the islands (Ghosh 2021, p. 98). Through the murderous activities of colonists and the clinging nomenclature used by scientists worldwide, Ghosh argues that much Indigenous knowledge of plants, forests, waterways and geologies has been lost. Acknowledgment of Indigenous knowledge still faces many problems of ongoing colonialism in interactions with Western science. Settler climate writers write to be respectful of already existing Indigenous knowledges of earthly entanglements. Indigenous writers' connections to Indigenous knowledges puts them ahead in the game. As Zoe Todd writes, 'Indigenous artists offer an important perspective on the inter-twined and relational connections between people and land, and through their art can craft concrete responses to the mess and violence of the economies operating in the Anthropocene' (2015, p. 252).

Indigenous knowledge is often not recognised until legitimated by traditional science (Nakata 2002). In many recent contexts 'scientific' proof of what is embedded in ancient cultural knowledge and stories is coming to light. Examples include meteorological knowledges, Indigenous readings of the stars, Indigenous architecture, Indigenous water management and Indigenous genealogies.[2] Groundwater narratives endeavour to balance Western science knowledge of groundwater with recognition of colonial devastation of not only many springs, but most significantly the cultural stories and beliefs that inherently give vitality to aquifers. For example, the Arabana people whose Country lies west of Kati Thanda (Lake Eyre) in South Australia have several creation stories related to how serpents created the mound springs (Department for Environment and Heritage, n.d.). The writhing serpent in the bubbling mound springs pools brings imaginations to life. Meaningful collaborations enable respectful enmeshments of Indigenous knowledge.

Opportunities for true collaborations that give primacy to Indigenous knowledge of the Elders however are rare (Hamacher and Anderson 2022). There is a growing movement to have Indigenous water rights acknowledged in water management and allocations. (Moggridge 2010; Marshall 2017; Hartwig *et al.* 2020). Interactions between hydrologists and Indigenous peoples' knowledge and custodianship of water bodies in an Australian context are particularly significant. Incorporating narrative groundwater's Indigenous flow lines asks climate writers to respect protocols for working collaboratively with holders of Indigenous knowledge.[3] Excellent examples of successful collaborations include the reciprocation between white scholars and Indigenous custodians and Bawaka Country to create the 'Songspirals' projects (Bawaka Country *et al.* 2022).[4] This project 'nourishes and shares Yolŋu understandings of song spirals to enable, enrich and awaken Country; to challenge and expand Western academic frameworks; and contribute towards more responsive relationships between people and places' (Bawaka Country *et al.* 2022, p. 435). There is a resonance between the ripples of songspirals and the cycling of science narratives as both circle and spiral around knowledge making.

Uncertainty

Hydrogeologists make many graphical and pictorial representations of groundwater based on data collected from monitoring groundwater flows through physical bores, seismic imaging and geological modelling. Uncertainty is however well acknowledged in the intricacies of the ways that groundwater is represented.

> Actual groundwater flow systems have mind-boggling complexity. The subsurface is a complex distribution of materials with transient fluxes of water. No matter how much effort is spent drilling, sampling

> and testing the subsurface, only a small fraction of it is ever sampled or tested hydraulically. The available data will provide only an incomplete picture of the actual system. Because of the inherent difficulty of characterising subsurface regions, substantial uncertainty is always introduced when creating the conceptual system.
>
> Fitts 2013, p. 235

Scientific papers in the hydrogeology field often acknowledge that many hydrogeological 'facts' are blurred and estimated. There is potential for conflicting interpretations of data. As data emerge about the connections between surface environmental flows, ecosystem maintenance and groundwater reserves, competition for use of limited groundwater is increasingly tested. For example, agricultural irrigators may be reluctant to hear the data, managed and monitored by complex bureaucracies, that suggest that 'their' water supply from aquifers is connected to river and forest maintenance possibly many kilometres away. Hydrogeologist from Oregon State University, Todd Jarvis (2014, p. 18), with his long career in negotiating settlements of groundwater conflicts around the world, points to the 'hydroschizophrenia' where separate surface water and groundwater governance authorities have differing regulations despite the knowledge of hydraulic connectivity between both surface and subterranean regimes. A climate writer's task in their research is to illustrate both the necessity for science and jurisprudence alongside the incumbent critique of scientific knowledge and legal policy frameworks. Building stories from conflicting viewpoints to illustrate groundwater's features and sustainability is an essential task in the creating of groundwater narratives.

While much of groundwater science is contested, the holy triumvirate in hydrogeology consists indisputably of three components: rock, space and water. After that, uncertainty and debate over meanings, concepts and viable extraction rates begin. Groundwater moves incredibly slowly through the minuscule pores in sandstone, drains through hairline fractures in basalts, seeps and drips through dissolved fissures in limestone. In some areas it follows the gravelly pathways of ancient rivers, called deep leads, long since buried by hundreds of metres of rock, sands and soils. Groundwater's flows are mesh-like and are determined by many features of deep geology. The scope for uncertainty and disagreements is enormous.

The 'capillary fringe', a thin or sometimes thick saturated layer just above the water table (Fitts 2013, p. 8), is a hydrogeological term inferring that the lines that designate the water table may be indistinct, the boundaries indefinite. A fringe has movement and aliveness. Another hydrogeological concept that speaks of uncertainties and sparks imaginations is the hyporheic zone. This defines the nebulous area beneath the beds of creeks, rivers and lakes where water moves between the water course or lake and the aquifers below. The pebbly or silty bottoms of rivers, creeks and lakes leak or gain water in an exchange with aquifers in plume-like flows. Rivers leak. They are not confined to their wandering paths across the surface. Topography, geological foundations, rainfall variations and the fluctuating movements of the water table all affect which direction waters move through the hyporheic zone. Flux, porosity and permeability (as discussed in the next two chapters) likewise imply the variability of water's subterranean actions within various groundwater systems. These terms and concepts attract metaphor and imagination.

Writing processes involve working through the unknowns to decisions on narrative elements such as structure, point of view and characterisation. From uncertainty and experimentation, climate writers create a form and style of writing that is relevant to the topics and questions being explored. Writers come to some slippery grip with how we think about groundwater's beingness as real matter. We also seek out contemporary, verified knowledge about what groundwater is 'doing'. Groundwater behaves like water. It is water, with all its complex molecular behaviours. It circulates as part of the global hydrological cycle; it at some stage becomes a vital component of human, nonhuman animal and plant lives. I use the term 'lives' here in the conventional meaning of being alive. As discussed in Chapter 2, the 'life', liveliness or vibrancy of matter, such as the vital forces in groundwater, that lead us to appreciate our enmeshments with its trajectories, are important to creativity, to imagining groundwater stories. Ultimately, the many uncertainties of the writing process and giving expression to groundwater's flows in many ways mirrors the uncertainties of hydrogeology.

Groundwater systems, meaning extensive basins with groups of interlinked aquifers, can dwell beneath thousands of square kilometres, with diverse rock types. Groundwater basins may also be relatively small and shallow. Surface landforms and deep geologies in the planet's crust affect the scale and temporality of groundwater volumes and flows. For example, the Great Artesian Basin, the largest freshwater groundwater system in the world, occupies 1.7 million square kilometres and lies beneath just over 20 per cent of Australia's landmass (Geoscience Australia n.d.). It supports massive ecosystems and over two thousand rural communities across inland Queensland, New South Wales and South Australia. Since colonial times, it has been severely depleted by excessive and unregulated bore drilling. Only since the 1970s have the effects of unregulated boring and waste of groundwater been recognised. Uncapped and freely flowing bores and evaporation and leakage from open water channels that move the artesian water from paddock to paddock have contributed significantly to lowering the water table across the Great Artesian Basin. A massive government-funded exercise to cap bores and close open water channels over the past twenty or more years is showing some sign of reducing profligate waste of groundwater to evaporation. Following recent wet seasons in eastern Australia (La Niña effects in 2021 and 2022), and above average rainfalls across the recharge zones to the Great Artesian Basin (which lies predominantly along the west side of the Great Dividing Range), new springs are emerging in unexpected places (McCosker 2022). It is unknown to what degree the Great Artesian Basin and which aquifers within this enormous system will benefit from rainfall, mostly because of the slow infiltration and recharge rates, and enormous distances the subterranean water flows. In some circumstances groundwater might move at about one metre per year. Nevertheless, increasing groundwater recharge has the potential to bring life and new micro- and macro-invertebrate species to some regions of central New South Wales, as the water table rises and new and old springs emerge.

Groundwater's subterranean flows can be measured mathematically by applying the elements of Darcy's law, a foundational calculation used by hydrogeologists to gain understanding of groundwater's subterranean movements. Henri Darcy was a Frenchman from Dijon, who in 1856 devised a mathematical formula to describe the rate of flow of water through a sandy substrate in a fixed volume, in a single direction. This was part of his work as an engineer, charged by local authorities with the responsibility to ensure a groundwater supply for Dijon. In its simplest form the formula states that water flow (Q) equals a calculation of permeability of the sand or rock (K), times gravitational head (I), times the area through which the water moves (A):

$$Q = KIA$$

There are numerous complicating factors that may be given a numerical value – such as variations in hydraulic conductivity of different rock types, the velocity of water flow over various distances, and of course the fact that water moves in multidimensional plumes (Fitts 2013, pp. 47–51). Mathematical formulae to capture the hydrological drivers of groundwater's transience involve complex modelling of a range of factors including time, climate cycles and geological variations over large areas (Currell *et al.* 2016). Efforts to trace aquifers' variable response times take account of changes in a region's water balance. Changes to how surface water and groundwater move may be caused by human-instigated land-use practices, such as constructing cities and highways, removing forests, changing deep-rooted grasslands to introduced pastures. The effects of climate change, such as sea-level changes and increasing prevalence of drought and floods, are also affecting groundwater (Novicky *et al.* 2010). Hydrogeologists seek information on aquifers' flow rates to trace, for example, drainage of pollution through leeching of herbicides and pesticides, petroleum fuels, or fire retardants. The complexities of factors that are built into mathematical models that trace such flows may be overwhelming for those untrained in the field.

Scientific understandings of groundwater are crucial to knowing this fluid and fragile 'character' in creative climate writing. In her list of proposals for authors of Anthropocene fiction, Lucy Sussex (2018) suggests that writers embrace science, 'take joy from the efforts of scientists, and learn from them'. I have learned immensely from the extraordinary work of hydrogeologists and, despite often feeling daunted by groundwater's complexity, found joy in the processes of learning. The science of hydrogeology, in combination with knowledge about the legislative arena and advances in engineering of groundwater drilling, are each necessary footholds in creating groundwater narratives. In my ongoing efforts to expand groundwater narratives, I continue to learn how aquifers form and flow, and the increasingly important relationship between hydrogeology and groundwater ecologies. The more science I learn, the more I am able to thread this 'behaviour' of groundwater into the narrative.

It is also incumbent on climate writers to learn introductory aspects of water law, policy and management. In Australia and internationally, this is a contentious, complex and shifting body of legislation and regulation, which is different in various jurisdictions and states. In Australia requirements within the Commonwealth's *Water Act* 2007, and its subsequent amendments, which aim for water usage to reflect sustainable environmental flows, often contradict state guidelines, which tend to prioritise local agricultural needs. The National Water Initiative, which emerged from the abolished National Water Commission in 2004–2005, introduced reforms to Australian water law following the Millennial Drought of 2000–2010. This initiative attempted to, but has not successfully, improved national consistency in sustainable surface water and groundwater management (White and Nelson 2018). Buying and selling surface water was legally 'unbundled' from the land where it lay or flowed. A free and unregulated water market now had ascendancy. The new rules favoured the viewpoint that water is a commodity to be used for profit by those who can afford to buy it (Hamilton and Kells 2021, pp. 53–56). Water is bought and sold by water traders, often with no connections to agriculture, meaning that prices increase and many farmers can no longer afford irrigation licences. The cost of water in a scarcity model becomes prohibitive. Access to water remains contested as water is considered a commercial commodity by some, an essential requirement for the survival of specific environments by others, and an essential connection to Indigenous Country by others. Contests for the use of groundwater follow a similar trajectory.

Science and Literature

Climate fictions are being written and read at a unique time, when anthropogenic impacts on the planet are gaining more and more attention. Science is trying to keep up with the extent of climate change investigations needed in the wake of new global atmospheric and meteorological variations and the devastations of ecological collapses. Literatures too are attempting to keep up with contemporary and predicted changes. As Charlotte Sleigh neatly sums up, 'Literature is how we decide what matters to us, and what matters will determine nothing less than the future of science' (2011, p. 24).

According to Charlotte Sleigh (2011), the common ground between science and literature includes: reliance on metaphoric and literal language; the use of subjective representations to present knowledge to readers; and the importance of passionate curiosity as a motivation for intensive inquiry. Regular dialogue between science and literature builds concurrence.

Sociological and historical studies of how scientists make and curate scientific knowledge have shown that the facts that emerge from science are not independent of their representation (Latour and Woolgar 1986; Clarke and Rossini 2011; Sleigh 2011, p. 16). The historical relationship between scientific knowledge and literature and readers' receptions of science's representations has changed over time, from being seen as anathema to religious doctrines, to scepticism, to blind acceptance (Sleigh 2011). Differing degrees of trust or disbelief in science have influenced how literature has portrayed science knowledge. Contemporary climate fiction writers may portray climate sceptics, climate activists or climate change refugees and survivors. Any or all of these various relationships to science knowledge may appear in the same story, the author aiming to portray anything from hope to despair in human responses to climate calamities. Readers cannot make sense of 'objective' facts without science researchers making subjective representations, such as reports, images, models or metaphors (Sleigh 2011, p. 4). Knowledge sharing and meaning making is affected by how the scientist represents their research findings. Incomprehensible language in little-known specialist journals accessible only to experts has been the norm, but is thankfully changing. The shared ground of seeking to make meaning in the spaces between hydrogeology and climate writing is more important than ever.

As well as learning of groundwater's existence and variable movements, my fictional work, as exemplified in the following excerpt, has developed an avenue to convey the anthropogenic causes of groundwater's problems and groundwater's responses to excessive extraction. Understanding human actions and relationships with groundwater, such as pumping bores or establishing laws to govern and supposedly control its course, is one way climate writers may present groundwater narratives. Embedding 'facts' or legislation at the same time as expressing groundwater's agency is the intention.

In the following excerpt from near the end of Chapter 4 of *Why we cry*, protagonist Frankie meets Helen, a hydrogeologist working for a government water authority. Ideas about what groundwater is and how it is affected by human actions are introduced. The excerpt touches on several minor characters and themes from the manuscript. Frankie is learning the value of communicating with experts to gain information, but sees the limits to her understanding of groundwater's potency in her life. The scene focuses on knowing the science of groundwater.

Excerpt 4.1: *Why we cry*, Chapter 4

Helen Mack overflowed her chair in her cubicle. A cascade of papers and empty coffee cups spread across her desk. The half-height velcro wall to her workstation was pinned with curling photos of distant hills and post-it tabs with cryptic reminders of tasks that may or may not have been achieved. Her desk was strewn with coloured maps and government reports, and she had two computer screens running, one connected to her laptop, the other to the departmental network, Frankie presumed.

Helen was in her early forties, square-faced, with dark currant eyes amid doughy skin that creased, delta-like, around her smile. She wore a baseball cap backwards, her straight nut-brown fringe pinned beneath the band across her forehead.

'Working in a bureaucracy, everyone's out to step on your neck, if they're not getting ready to stab you in the back', she whispered. 'The hat sort of makes them keep away.' Helen wore striped leggings over which a long floral chemise hung from her bulging chest. She wore leather-strapped sandals. Her Blundstone boots were under her desk, for fieldwork days, Frankie thought. Helen worked for Central Goldfields Water Authority, monitoring groundwater pollution across a region stretching from the Murray River to the northern edges of Melbourne's metropolitan sprawl. It was her first job out of uni, after a string of degrees, Clem had told Frankie. She'd stayed.

'Thanks for seeing me.' Frankie half stood to shift in her seat.

'Fire away.'

Frankie had written down some questions, remembered the first one, but kept the list in the pocket of her jeans.

'If water is everywhere, constantly recycling itself, why is it so scarce? And humans are made up of it, aren't we?'

'Seventy per cent water in most mammals. Can't comment on human biology, that's not my field. Ask me about hydrogeology.' Helen wiped a bead of sweat from her top lip, looked at Frankie, fierce and puzzled and kind all at once. She continued before Frankie could speak. 'Water cycles, recycles itself, globally. Water is finite. The same water that appeared somewhere four and a half billion years ago might now come out of our taps, flow in our rivers, seep into our aquifers. That's where it rests, away from the hurly-burly of surface water activity.' Helen moved her cushion-like arms in a wave, as if she couldn't resist the languor of groundwater. Frankie noticed her use of the possessive. Whose taps? Whose aquifers?

'You're saying that water's lucky to get underground. It's unlucky to be sucked up prematurely.' Frankie still thought of drilling for ancient water as like aborting something before its time is due. She'd also read about something called the hydrosocial cycle, but was too intimidated to ask Helen what she thought about that.

'It's unlucky to collect the pollution humans keep dropping into it. Domestic sewers, industry wastes, big and small. Petrol station leakages, dry cleaners releasing solvents into stormwater, agricultural pesticides, don't get me started on coal miners and their deep holes.' She pushed her cap to the back of her head. 'Let alone chemicals used in fracking for coal-seam gas, fire retardants at airports.' She puffed, drew breath.

Frankie took a small notepad from her backpack, made some notes of things to follow up. What was it about dry cleaners and groundwater?

'There are dead water zones, where groundwater is sort of killed by age, salt. But that's different to chemicals, pollution. It's people making pollution. DDT, you know, Rachel Carson's book? *Silent spring*.' Helen looked into the dregs of a couple of empty coffee cups. 'Sometimes the age of the aquifer marks the tragedy. Water can be down there for tens of thousands of years.' Helen kept throwing ideas, giving no space for Frankie to respond or digest.

Helen explained the effect of living in an area with a gold history. 'Groundwater has got a white-fella story around here. And a black-fella story too.' She sipped the milky remains of a cold coffee. Winced. 'Below Cawling and Broadville, gold mining tunnels, honeycombed in parts. They bucketed out the water inflows.' Helen was on a roll. 'The first recorded water trading was from a guy in the early 1850s who was pumping water from his shafts, sold it, in a drought summer, to another guy downstream who was sluicing, needed something to wash his slurry when the creek stopped flowing over summer.'

Frankie knew that capping off old mineshafts was superficial. Many mines and disused wells in the bushland surrounding Cawling still gaped open, leaking dank air from dark underground passages. Her father, Lennie, had often said to Frankie to watch out walking in the bush around their place. She remembered his regular beef. 'No one thinks below the surface, unless it's about making money. Especially Kasha Duff', he'd added. 'He'd drill, scrape, excavate a hole even if it was for nothing. He thinks he's having an impact by making a quid In his own short life.' Frankie remembered her father's hands rubbing his chin.

Helen coughed. Frankie muttered, 'Sorry.' She'd had only partly applied herself to the first-year geology subject, years ago. It was all complicated terms for incomprehensible time periods and rock-type names were like another language. She wished she'd listened more. Helen galloped on, describing local geological events, granite uplifts and exfoliation, basalt flows, how slate is metamorphosed into shale. Ordovician, Quaternary epochs did something. Frankie scrawled notes, later indecipherable.

'And then there's the law.' Helen punctuated her monologue, wiped her brow and sipped her cold coffee again. Sucked in a breath.

Frankie sat forward to speak. 'The law says that people can only own the top six inches when they buy land, doesn't it? That means all the rest is open to prospecting, right?' Frankie had read rudiments of the legal situation, checked now with Helen.

'No and yes. In the USA you own to the centre of the core under your land. In our country, you own the land fifty feet below, within your boundaries, just over fifteen metres. Any minerals, stone and water, though, belong to the Crown. But you can get royalties. The Crown gives out licences for groundwater.' Helen tapped a report on her desk. 'But regulation is the missing ingredient. And monitoring.'

'That's incredible. Who's the Crown?'

'Hypothetically, water and minerals are the property of the Crown', Helen paused. 'Who is the Crown?' She repeated the question. 'The government, well sort of. The government agencies allocate the licences to mine. Yes, mining water happens. In reality humans think of it as "ours" when they collect it. In dams and tanks.'

'All water? Rain? Or groundwater?'

'It's all mixed up in the rules and reality.' She paused, sighed, swatting a fly away from her coffee mug. 'To get a groundwater licence, for extraction for commercial gain, you have to apply to us, the water authority, first. Comes through the department, upstairs. Licence to drill and construct. It's complicated.' Helen paused. 'Some of the clunkiest laws this country hangs on to.' Helen looked around her office, over the cubical wall, as if nervous that a colleague might challenge or report her for making heretical comments. 'They don't challenge the laws of the country in this department. We just execute them', she whispered behind her pudgy hand. 'I'm trying to tell them upstairs that the EPBC is relevant to our work.'

'EB ... ?'

'Environmental Protection and Biodiversity Conservation Act. Legislation passed by government when they had some sense. It's about GDEs. Groundwater-dependent ecosystems.' Helen paused. Frankie scratched at her notes. 'Looks like this is getting me into deep water, so to speak.' Helen was glancing over her shoulder again. 'Maps create a place for places to speak.'

'Did you say maps are a place for voices to speak?' Frankie asked.

'Places. Voices. They all speak. I'll show you the 3-D images next time.' Helen turned to her computer, stared at the long queue of emails awaiting her attention.

'This is like, another language.' Frankie felt like she was swimming, just holding her head above water.

'Sorry, listen, I gotta go. I've got a tight timeline to meet. I'm assessing a cemetery application, for a new bore. They want to keep the lawns green, and the ponds full in the Vietnamese section. Reflection and that. Cultural stuff. I have to work out how much seepage there'll be.'

'Sure, thanks.' Frankie turned to leave. 'How about a meal with me and Clem sometime?' She felt herself flush at her temerity. She'd only just met the woman.

'OK. Great. Give me a call.'

Frankie found her way out through the confusion of aisles and workstations, each occupied cubicle with its own fragments of decorations. Conundrums clashed. Pumping underground water to green up a cemetery, recycling life amidst death. She re-entered bright sunshine outside the office block, blinking.

Three weeks later, Helen and Clem sat around the table at Frankie's. The kitchen was thick with stewed aromas. Red wine bottles breathed, glasses clinked, conversations started with little warm-up.

'What do you think of the term Anthropocene?' Clem asked Helen, thinking that she'd give a tight geological argument off pat.

'Some geologists think that humans are altering the planet, now more than ever in the past, that the effects of human activities can be seen in geological processes. Some don't.'

'Haven't humans always affected where we live, changed things, say with fire?' Frankie asked. She felt her skin prickle.

'Yes, but that mightn't make a geological epoch.' Clem stood to stir the pot on the stove, inhaled deeply over the savoury vapours, gently replaced the lid.

'Some like the word Pyrocene. We've always been burners.' Helen said. 'This level of human burning of carbon into the atmosphere is something bigger.'

'If global warming is about increasing carbon in the air, how does that affect geology?' Frankie asked.

'Some say the planet is cooling, agelessly cooling after the first big bang. That this blimp, this spike on the graphs, is nothing in the long term.' Helen swirled her wine glass.

'You geologists always take the long-term view. Super long.' Clem egged them on. 'Least my little mitochondria are immediate.' She paused. 'Well sort of.'

'I'm not saying that there's not atmospheric warming, and oceanic warming, and acidification.' Helen pushed back her fringe. 'Ecological disasters, extinctions. And we could say that the thin layer of carbon, from us burning fossil fuels for only two centuries, is making a hell a lot of damage. There's plenty of evidence and I believe the science. It's just that the term climate change, maybe even global warming now, carries more political fog than is useful. Anthropocene has got problems too.'

'What do you mean? Doesn't the environment need political clout to get anything done about this mess?' Frankie started to ladle the vegetable casserole into bowls.

'Yes, sure. But take, for example, today. I had to write a submission for funding to assess the risk of the Murray-Darling Basin Plan on groundwater systems in this region. We have to revise the groundwater management plans around here, as water usage around the Murray changes. How do we fit that job in? In the submission I had to show predictions of weather patterns for 2050, only 25 years away. It was near on impossible. They think we know.'

They each savoured the meal, dunked thick slices of bread.

'Not even the most canny of scientists really know. We're the first to admit blindness', Clem said, chewing.

'The story of groundwater is the story of geology, a story that stretches into deep time', Helen continued. 'And we never ...'

'Never say never', Frankie piped up, hearing her mother's voice in her own.

'Well, we never really see it, while at the same time I'm mostly made of something like it – water.' Helen stared into her bowl, as if searching for the answer to a philosophical question.

'I use a microscope to see my subjects. My prosthetic eye.' Clem held her circled fingers to her eyes, rolling them in a Frankenstein sort of way.

'Different stories in different places.' Frankie said between mouthfuls.

'Geology tells different stories to different people. Water is what you make of it. It all depends on where the water flows, and how. Groundwater is not the same everywhere. It has to occupy the space it has.' Helen nodded at her plate, thinking. 'It flows through different rocks, different formations, in different ways.'

'Maybe we need to turn to artists.' Frankie waved her fork, like an antenna seeking ideas, a painter flourishing a brush.

'So it's up to artists to explore shapes, sombre moods, movement. How would they paint groundwater's character?' Clem was wide-eyed. 'Impossible', she signed.

'Yeah, why not? Watercolour runs.' Frankie remembered pigments seeping into fibrous pores of thick, wet paper from her high school art classes. 'Uncontrollable.'

'I put my faith in evidence, hard words, hard data', Helen said, shaking her head.

'Just thinking of options.' Frankie muttered.

'We know more than we did in the past.' Clem paused, raised her glass. 'To the cook.'

'But we still don't know enough.' Frankie clicked her empty bowl with her spoon.

In central Victoria, Australia, where the narrative *Why we cry* is set, groundwater occurs in five main aquifer types: Ordovician sedimentary bedrock; granite; gravelly deep leads; basalt from old larva flows; and shallow alluvial aquifers. Deep leads in this region are unconsolidated sands and gravels of ancestral streams buried approximately sixty or more metres beneath ancient basalt flows that originated from volcanoes around twenty million years ago (Goulburn Murray Water 2013). The geology of the district and how it affects groundwater flows is interwoven into the narrative.

Science Communication

Science communication often takes the forms of reports, government papers and journalism. Various social media and numerous websites have progressed the dissemination of climate change information effectively, yet few make forays into the perils and potency of groundwater[5] (White and Nelson 2018; Nelson *et al.* 2020). My aim in this chapter is not to discuss the technicalities of communicating complex hydrological science. The large literature on *science communications* is beyond the scope of the book (Leßmöllmann *et al.* 2019; Illingworth 2020). It is often journalists who mediate between science knowledge and public discussion of climate change (Baucom and Omelsky 2017, p. 7). Public interest journalism focuses on dissemination of current and verifiable information. It faces the threats from fake news and unverified opinions masquerading as 'fact'. Needless to say, making elementary aspects of hydrogeology accessible, through providing comprehensible science information in any of the creative narrative forms (fiction, poetry, screen and play scripts, creative non-fiction, etc.), has elements of science communication inherently embedded. For creative climate writers the challenge goes beyond communicating what might be current groundwater policy and information about groundwater's features and actions, or reporting newsworthy incidents or stories of groundwater loss, its causes and consequences.

As Sam Illingworth says:

> Science is an essential tool for the development of a healthy, informed, sustainable, and balanced society. It is also a tool that has an unfortunate history of being used to ostracise, alienate, and exclude; a weapon of choice for those who wish to maintain the status quo of a western, patriarchal hegemony of thought.
>
> Illingworth 2020, p. 1

Accessible literatures have an essential role to inform and move readers who may otherwise be excluded from complex science communications.

Illingworth (2020, p. 1) proposes the use of poetry as a potent means to assist scientists to engage in what he calls 'outward facing' communication to the wider community, as well as, to a lesser extent, in 'inward facing' communication within the science establishment. Poetics, he argues, offers 'a unique way to help critique, interrogate, and diversify science … Poetry has the capacity to hold a mirror to science, capturing its reflection warts and all' (Illingworth 2022, p. 3). Various forms of climate writing have a similar capacity. Hydrogeology is not a well-known science and its narration – in poetic or long-form prose – will broaden efforts to ensure groundwater's sustainability. Thinking of how to portray scientists and science information, creative writers are not merely attempting to disseminate accurate scientific information. Climate writers create opportunities for 'multi-directional dialogues' and collaborations between writers and scientists to broaden, deepen and diversify meaning making for wide readerships (Illingworth 2022, p. 4).

Conveying to readers how aquifers form and flow in various geological formations is a starting point. Increasingly important relationships between surface and subterranean waters, and between hydrogeology and groundwater ecologies[6] also underpin how climate writing engages with contemporary water problems (Humphreys, 2008). Foundational aspects of groundwater's actions and relationships comprise the fodder of subterranean imaginaries. Writers are called to imagine the recharge pathways of waters from subsurface unsaturated soils, through the capillary fringe to gradually seep into rock or stone layers where it merges with saturated aquifers and plumes slowly beneath landscapes. Aquifers interact with nearby rivers and streams, swamps and lakes, at times feeding subsurface flows into water bodies, at other times drawing water from the same bodies of water. How the aquifer provides water for deep-rooted trees, or for unique stygofauna, and for groundwater-dependent ecosystems is another part of the story. Some have argued that there is no fixed boundary between rationality and emotion (Latour 2018). A story wedged too tightly to science has the potential to lose cultural, metaphoric and mythic means of making meaning. Climate writers need to cross the boundaries between scientific rationality and literary expressions of emotion. Subterranean imaginaries take the 'facts' and enmesh them in human and nonhuman lives through the use of metaphor and narrative.

Science and Law

In the 1960s, Rachel Carson's *Silent spring* (1962) lamented human hubris, lost landscapes and the health threats posed by agricultural use of the pesticide dichlorodiphenyltrichloroethane, or DDT. Her book spurred community and eventually political outrage at the poisoning of not only plants and animals, but entire landscapes and communities. Carson's beautifully crafted polemic continues to be widely influential and is credited with founding the environment movement in the late 1960s. The book contributed to changes in legislation in the USA on water pollution and environmental protections. Its hybrid non-fictional form, irrefutable logic and careful presentation of evidence illustrates the capacity of writing to bridge scientific knowledge with transformative public opinions. Pesticides were used in agriculture in the aftermath of their deployment in the Second World War. At the time, Carson's proposition that these practices caused localised and landscape-wide environmental and health problems contradicted all government policy and standard agricultural practices. Following widespread debate of Carson's work, which included unsuccessful attempts by chemical companies to discredit both the book and Carson, DDT was banned in the USA in 1972 as one of the first actions of the US Environmental Protection Agency (Pesticide Action Network 2015). Despite scientific knowledge about the damages done to human and nonhuman ecologies, many beliefs about the use of herbicides and insecticides in industrial food production have however not changed significantly. The organic food industry provides for a small and wealthy proportion of consumers. Environmental battles continue to be fought over groundwater pollution, and increasingly over access to groundwater.

There are many site-specific examples writers can draw upon to fuel groundwater narratives based on contestation between both the science and the laws governing groundwater use and its local users and cultural custodians. The story of a localised dispute about groundwater usage in the small Victorian township of Stanley is a classic tale that portrays a common battleground. Tim Carey, from Stanley Pastoral, obtained licences to extract nineteen megalitres per year of groundwater (that's about nineteen Olympic-sized swimming pools each year) and sell it to a mineral water company for bottling (into plastic bottles) and distribution worldwide.

Through the effects of drawdown, a cone-like lowering of the water table around the very deep bore that Carey established, the local farming community faced potential loss of access to groundwater for agriculture. Orchardists, hobby farmers and local farmers joined forces to retain their access to groundwater. 'Stanley Water for Stanley' was the local community catchcry of those objecting to Carey's commercial project. The arguments over conflicting hydrological reports were fierce and were finally heard in the Supreme Court of Victoria in December 2016. Through a convoluted interpretation of Australian water law, rather than assessments of hydrogeological reports, Carey's licence to take the groundwater was upheld. Members of the Stanley Water for Stanley group were devastated, none more than the organic fruit-growing neighbours who now have water trucks carrying groundwater away along the dirt roads that go past their front gates. Caught adrift between competing hydrogeologists' reports and ambiguous laws, Stanley residents and farmers wonder how long their bores will operate, and how local waterways, which directly receive groundwater when the water table is high, will be affected. Locals fear that taking excessive groundwater for this commercial venture will permanently reduce both groundwater and surface water flows.

Mining groundwater is the term for extraction of groundwater beyond the recharge capacities of the aquifer. Slow-flowing aquifers may take centuries to recharge; some will never be replenished. Water wars and complex legal wrangles are predicted to become more prevalent (Jarvis 2014). Understanding the legislative context of how policy makers, farmers and property developers have used and mined groundwater as if it is an endless resource enhances the potency of groundwater narratives. Enmeshing groundwater narratives with underpinning scientific evidence is important to challenge commercially driven exploitation of resources that are both unsustainable and potentially damaging to communities and ecosystems.

Presenting credible science and contemporary legal frameworks about groundwater is not only for peace of mind, but it also garners readers' trust in the author's accuracy (Webb and Brien 2011, p. 197). Many issues are still debated within hydrogeology as new evidence is found. Science is constantly questioning itself. Water policy and law is updated frequently, under different political contexts and legislatures. Changes and contests become incorporated in creative works. Groundwater narratives based on scientific research aim to give readers confidence in the author's fidelity to science and law, including clear pictures of the conundrums. Climate writers aim to find 'harmonies and synergies', where fact and fiction are combined to make new narrative and scholarly connections that generate the 'wider implications and applications' of the author's research (Webb and Brien 2011, p. 199). A valuable strategy in confirming a writer's application of science and law is to ask hydrogeologists and water lawyers to read drafts of relevant scenes. It is my experience that consolidating collaborative relationships between writers, scientists and water lawyers bears fruit. For example, a water lawyer working with Indigenous communities on water rights guided me through the legal processes of water licence appeals as evoked in the following excerpt. Accessing groundwater in Victoria requires a licence to construct a bore under the Victorian State government's *Water Act 1989*. My interest in conveying an accurate depiction of groundwater licensing processes is as strong as my interest in conveying legitimate science.

Excerpt 4.2: *Why we cry*, Chapter 11

A fortnight after her last exam, the hearing in the Victorian Civil and Administrative Tribunal loomed. Save Our Groundwater had raised just enough funds to go to VCAT on two counts: to ask it to revoke the take and use licence for Kasha's bore, and to object to the council's permit allowing him to construct the bore site for the twenty-five houses. Evidence that he'd already started pumping was impossible to confirm. They aimed to halt his plans to extract more water. Bruce Landy, their friend the water lawyer, helped them prepare the case, and stood with them on the first day of the hearing. Lois spoke on behalf of the group, listing problems of water shortages for the community, arguing, with some coaching and data from Helen and Frankie, that increasing groundwater extraction was not sustainable and would rob too many bores in the district. Clem provided her with the names of unique stygofauna that lived in the liminal zones of the river and in sub-surface soils, microfauna that acted like krill in the ocean as the foundation of ecosystem health. Frankie watched Lois's knee jiggle as she sat through the hearing. They listened to Colin Fitts from the water authority repeat the arguments about complying with the local groundwater management plan. Bruce Landy leapt to the podium, questioned the currency of the plan, pointing out that the ten-year-old data did not include consideration of new information on the effects of climate change. At the end of the day a representative from the local council argued that the advantages of attracting new ratepayers to the district outweighed all else, and that they were entitled to follow the water authority's recommendation. If there really was enough water, Frankie mused.

Lois, Clem and Frankie spent the evening in the city, booked into a motel near the VCAT offices. They wanted to hear the commissioner's summary the next day. Too far to make two train trips. City lights, traffic drone and sirens' wails hadn't changed.

They arrived at the VCAT offices early the next day for the summing up. Frankie was surprised to see Wade in the foyer standing close to three other older people. Silver hair and loose-fitting suits.

'Didn't know you were coming down', Frankie said. She nodded to his compatriots, recognising Wade's friends as those she'd seen at Bruce Landy's office.

'There's quite a bit I couldn't tell you. It's out now. Our turn.' Wade shuffled, turning a circle with the toe of his boot. 'Bruce Landy helped us too.'

'What do you mean?' Clem edged in.

'Yaluk Cooperative are also contesting Kasha's license. Different grounds. Today.'

'Why didn't you tell us?' Frankie fiddled with a plait of grass she'd made, found in her pocket.

'Couldn't. Bruce reckoned that the legislation was too untested. And it had to be Yaluk voice, no interruptions.'

'That's great.' Frankie followed as they filed into the tribunal.

Wade fiddled with his collar and tie as his Auntie and Uncles addressed the tribunal. As they each spoke Frankie imagined spirits seeping through the wood-lined walls, filling the air with story. Ancestral voices nourished them. Frankie listened closely as Yaluk members spoke of generational connections to Country. Their voices wavered at times, but their ancestors lifted them with gravitational solemnity that held the VCAT tribunal room transfixed.

Wade stood when it was his turn to speak.

'Please explain exactly what are your cultural links to groundwater.' The pink-faced lawyer hearing the tribunal peered over rimless glasses. Clem grabbed Frankie's hand; Lois clutched the other.

'You could call it dinosaur water, your honour, and you wouldn't be too far wrong. Some of this groundwater that so many people wash in was laid on the surface when dinosaurs were around. It seeped fathoms and flowed thousands of kilometres. This old water is connected to long time past times, to awe-inspiring epochs. Our old stories come from those times. Our old stories tell some of what we know. We drink this water and drink time.' Wade spoke on, talking of grandparents and great-grandparents killed in flu epidemics, of murders and massacres, of washing blood in river pools. He referred to his compatriots with respect, backing up their descriptions of Yaluk Cooperative's decade-long claims for acknowledgment of connection to Country, their descriptions of scar trees and middens that line the waterways, evidence of long, long occupation and custodianship of Country. 'It's not just the rivers, it's what's under the rivers that counts. That's where the water must be protected. The rivers are dying because too much water is being taken. Taken.' He repeated the word and everyone knew the history of theft that the word referred to. 'It's not just the rivers that are our cultural lifeblood, it's all that surrounds them, all that lies beneath. We were not asked. The licence to drill for our water is not valid because we were not asked. Groundwater is as important for culture, as the air we breathe.' He breathed out, letting go the speech he'd rehearsed, sighing for the power of what he'd learned. In finding words and letting them permeate he appeared to Frankie to grow taller, to emanate strength.

It would be three weeks before the outcome of the VCAT application would be announced.

The fictional universe I created in the manuscript *Why we cry* enmeshes close relationships between what hydrogeologists currently know about groundwater, the current legislative framework, and the social and environmental context in which the fictional human actors play out the story. My characterisation of Frankie Pankhurst focuses on her internal and external conflicts and the emotions generated by her actions in the face of loss of groundwater. It also conveys her connections to community members who support her learning about groundwater law and science. Her journey towards environmental activism is a consistent thread. Her learnings about groundwater mirrored mine at the time of writing.

Climate writers make decisions on genre, style or form, bridging between the known scientific facts and the necessary imagination. Groundwater's characteristics and affects are illuminated through conversations between fictional human characters, though a character reading a newspaper or by presenting the findings of a hydrogeological report, or in the words of a hydrogeologist. The proximity of science and law may be comfortable or bristling with conflict in various scenarios.

Groundwater may be lugubrious in its movements, slow in the ways it changes chemical constituency, absorbing and depositing salts as it moves through layers of rock. But being slow does not make it passive. Groundwater narratives create scenarios where scientific knowledge and artistic, imaginary ideas merge, melt, bleed together with the politics and effects of human activities. Blending the knowns of science with the intangibles of human affect to create credible stories becomes a climate writer's haven. The entelechy, the vital force, of water's subterranean actions, need not necessarily be seen to be believable. Hydrogeologists know something of groundwater's movements beneath a wavering water table. It is an underground affair, into which groundwater narratives flow.

The next chapter explores how new nature writing evokes subterranean places.

Notes

1 Ursula Le Guin builds the idea of narrative as 'carrier bag', which is filled with beginnings, processes and transformations, as Lesley Instone explains in the essay 'Writing difference differently' (Fisher *et al.* 2015). Narrative as carrier bag acknowledges the porosity of story-making processes.

2 David Jones and Darryl Low-Choy (2021) provide an excellent overview of Indigenous knowledges and the recent links to Western knowledge in *Indigenous Knowledge Systems* and *Yurlendj-Nganjin*.

3 Attempts to integrate traditional and scientific knowledges face many problems, including diversity of knowledge holders, the sacred and secret nature of some knowledge, and constant negotiating of the cultural interface of knowledge systems (Nakata 2002). These discussions are beyond the scope of the book, but are essential for climate writers and groundwater scholars to engage with.

4 Veronica Strang (2013), Sarah Wright *et al.* (2012) and Brad Moggridge (2010) have also contributed greatly to collaborations that expand knowledges of Indigenous waters.

5 See the specialist webpages on Australian groundwater at Bureau of Meteorology (2022) and Visualising Victoria's Groundwater (2019) for examples of publicly accessible groundwater information.

6 See Humphreys (2008) for a full discussion of how the disciplines of hydrogeology and groundwater ecology link with and inform the other.

References

Baucom, I. and Omelsky, M., 2017. Knowledge in the age of climate change. *South Atlantic Quarterly*, 116, 1–18.

Bawaka Country *et al.*, 2022. Songspirals bring country into existence: singing more-than-human and relational creativity. *Qualitative Inquiry*, 28, 435–447.

Brodie, R., Lawrie, K. and Commander, P., 2012. Groundwater – lifeblood of the continent. *In:* R. Blewett, ed. *Shaping a nation: a geology of Australia.* Canberra: Geoscience Australia and ANU Press, 332–279.

Bureau of Meteorology, 2022. *Groundwater information* [online]. Available from: www.bom.gov.au/water/groundwater/ [accessed 20 January 2023].

Carson, R., 1962. *Silent spring*. Boston, MA: Houghton Mifflin.

Clarke, B. and Rossini, M., 2011. *The Routledge companion to literature and science.* London: Routledge.

Currell, M., Gleeson, T. and Dahlhaus, P., 2016. A new assessment framework for transience in hydrogeological systems. *Groundwater*, 54, 4–14.

Department for Environment and Heritage, n.d. *Strangways interpretation signs* [online]. Available from: https://cdn.environment.sa.gov.au/environment/images/strangways-information-PDF.pdf [accessed 19 February 2023].

Fisher, K., *et al.*, 2015. Writing difference differently. *New Zealand Geographer*, 71(1), 18–33.

Fitts, C.R., 2013. *Groundwater science.* Waltham, MA: Academic Press.

Geoscience Australia, n.d. *Great Artesian Basin* [online]. Available from: www.ga.gov.au/scientific-topics/water/groundwater/gab [accessed 18 August 2022].

Gerard, P., 2018. Creative nonfiction: researching and crafting stories of real life. Long Grove, IL: Waveland Press.

Ghosh, A., 2021. *The nutmeg's curse: parables for a planet in crisis.* London: John Murray Publishers (Hachette).

Goulburn Murray Water, 2013. Central Victorian Mineral Springs Groundwater Management Area: Local Management Plan. Tatura, Vic: Goulburn-Murray Water.

Hamacher, D. and Anderson, G.M., 2022. *The first astronomers: how Indigenous elders read the stars.* Crows Nest, NSW: Allen & Unwin.

Hamilton, S. and Kells, S., 2021. *Sold down the river: how robber barons and Wall Street traders cornered Australia's water market.* Melbourne: Text Publishing.

Haraway, D., 1988. Situated knowledges: the science question in feminism and the privilege of partial perspective. *Feminist Studies*, 14, 575–599.

Hartwig, L.D., Jackson, S. and Osborne, N., 2020. Trends in Aboriginal water ownership in New South Wales, Australia: the continuities between colonial and neoliberal forms of dispossession. *Land Use Policy*, 99, art. 104869.

Humphreys, W.F., 2008. Hydrogeology and groundwater ecology: does each inform the other? *Hydrogeology Journal: Special Edition – Hydrogeoecology*, 17, 5–21.

Illingworth, S., 2020. *Effective science communication: a practical guide to surviving as a scientist.* Bristol: IOP Publishing.

Illingworth, S., 2022. *Science communication through poetry.* Cham, Switzerland: Springer.

Jarvis, T., 2014. *Contesting hidden waters: conflict resolution for groundwater and aquifers.* New York: Routledge.

Jones, D. and Low-Choy, D., 2021. *Indigenous knowledge systems and Yurlendj-Nganjin.* Newcastle upon Tyne: Cambridge Scholars Publishing.

Latour, B., 2018. *Down to Earth: politics in the new climate regime.* Cambridge: Polity Press.

Latour, B. and Woolgar, S., 1986. *Laboratory life: the construction of scientific facts.* Princeton, NJ: Princeton University Press.

Le Guin, U., 1984. The author of the acacia seeds and other extracts from the Journal of the Association of Therolinguistics. *In:* U.K. Le Guin, *The compass rose.* London: Panther Books, 11–19.

Leßmöllmann, A., Dascal, M. and Gloning, T., 2019. *Science communication.* Berlin: De Gruyter Mouton.

Marshall, V., 2017. *Overturning aqua nullius: securing Aboriginal water rights.* Canberra: Aboriginal Studies Press.

McCosker, M., 2022. Across Outback Queensland, Great Artesian Basin water is surfacing where it has never been seen before. *ABC News – Rural*, 17 August [online]. Available from: www.abc.net.au/news/rural/2022-08-17/emerging-springs-across-queenslands-great-artesian-basin/101260220 [accessed 18 August 2022].

Moggridge, B., 2010. *Aboriginal knowledge and cultural values of water.* Canberra: CSIRO.

Nakata, M., 2002. Indigenous knowledge and the cultural interface: underlying issues at the intersection of knowledge and information systems. *IFLA Journal*, 28, 281–291.

Nelson, R., Barnett, S. and Kollmorgen, A., 2020. The evolution of groundwater management policy in the states of Australia. *In:* J.-D. Rinaudo *et al.*, ed. *Sustainable groundwater management.* Cham, Switzerland: Springer, 129–141.

Novicky, O., Kasparek, L. and Uhlik, J., 2010. Vulnerability of groundwater resources in different hydrogeological conditions to climate change. *In:* M. Taniguchi and I. Holman, eds. *Groundwater response to changing climate.* London: CRC Press, 1–10.

Pesticide Action Network, 2015. *The DDT story* [online]. Available from: www.panna.org/resources/ddt-story [accessed 22 August 2022].

Richard-Ferroudji, A. and Lassaube, G., 2020. The challenge of making groundwater visible: a review of communication approaches and tools in France. *In:* J.-D. Rinaudo *et al.*, eds. *Sustainable groundwater management: a comparative analysis of French and Australian policies and implications to other countries.* Cham, Switzerland: Springer, 191–209.

Rockstrom, J., *et al.*, 2023 *The what, why and how of the world water crisis: Global Commission on the Economics of Water phase 1 review and findings.* Global Commission on the Economics of Water. https://watercommission.org/wp-content/uploads/2023/03/Why-What-How-of-Water-Crisis-Web.pdf [accessed 18 March 2023].

Segal, M., 2017. The missing climate change narrative. *South Atlantic Quarterly*, 116, 121–128.

Sleigh, C., 2011. *Literature and science.* Basingstoke: Palgrave Macmillan.

Strang, V., 2013. Conceptual relations: water, ideologies and theoretical subversions. *In:* C. Chen, J. MacLeod and A. Neimanis, eds. *Thinking with water.* Montreal: McGill-Queens University Press, 185–211.

Sussex, L., 2018. The end of the world as we know it. *Sydney Review of Books*, 17 September [online]. Available from: https://sydneyreviewofbooks.com/dyschronia-jennifer-mills-how-to-bee/ [accessed 22 September 2018].

Todd, Z., 2015. Indigenizing the Anthropocene. *In:* H. Davis and E. Turpin, eds. *Art in the Anthropocene: encounters among aesthetics, politics, environments and epistemologies.* London: Open Humanities Press, 241–254.

Visualising Victoria's Groundwater, 2019. *Visualising Victoria's Groundwater* [online]. Available from: www.vvg.org.au/ [accessed 20 January 2023].

Webb, J. and Brien, D.L., 2011. Addressing the 'ancient quarrel': creative writing as research. *In:* M. Biggs and H. Karlsson, eds. *The Routledge companion to research in the arts.* Abingdon: Routledge, 186–203.

White, E.K. and Nelson, R.L., 2018. What happens to small towns whose water becomes big business for bottled brands? *The Conversation*, 17 May [online]. Available from: www.abc.net.au/news/2018-05-17/what-happens-to-small-towns-whose-water-becomes-big-business-fo/9770128 [accessed 5 July 2018].

Wright, S., *et al.*, 2012. Telling stories in, through and with Country: engaging with Indigenous and more-than-human methodologies at Bawaka, NE Australia. *Journal of Cultural Geography*, 29, 39–60.

5

POROSITY

PLACE AND NEW NATURE WRITING

Figure 5.1 Bill Sampson, *Two roads (part II)*, 2020, synthetic polymer on paper, 60 × 60 cm.

DOI: 10.4324/9781003270416-5

Rock and soil particles pack together in different ways to give the material various degrees of porosity. Porosity, or the open spaces within a geologic formation, is affected by the grain size, the similarity of grain sizes, the ways they pack together, and whether there are mortar-like substances in between the grains. For example, sandstone has largish crystals of various sizes, with big spaces between them, making it quite porous. Slates have grains of similar sizes that are closely packed together and so are less porous. Porosity affects the permeability of rock (the rates of flow of liquids). Higher porosity usually enables greater permeability, or higher flows of water. Porosity is calculated mathematically as a fraction or ratio of the total material volume that is pore space. That is, the ratio of the volume of the pore spaces between the soil or rock particles that can be filled with air, gases or liquids, and the total volume of a sample of the rock or soil. Porosity is expressed as a percentage. For example, many soils have porosity in the range of 40 to 50 per cent. Sandstones' porosity may be between 10 and 35 per cent, as the large irregular grains lead to large pore spaces. Granite, as a crystalline igneous rock, has low porosity, less than 1 per cent.

> In Yukatan, you never see the water. It flows underground, beneath a fragile sheath of earth and limestone. Occasionally that delicate Yukatan skin blossoms in eyes of water, in liquid pond – the *cenotes* – that attest to the existence of a mysterious subterranean current. For me, love is like those hidden rivers and unexpected streams of the Yukatan.
>
> Fuentes 2004, p. 1

The cenotes of Mexico's Yucatan Peninsula are water-filled sinkholes connected to fathoms-deep cave networks, where people swim with submersible headlights strapped above their goggles. Fuentes (2004) reminds us that writing of subterranean places is more than entering imagined or real 'liquid ponds'. Nature writing for centuries has expressed human appreciation of the natural world, awe or embeddedness in environments, seasons, habitats (Williams 1975). Emerging trends in new nature writing are responding to human and nonhuman entanglements across a warming planet. This chapter examines the politics and the epistemological critiques of representation in expressions of place in nature writing. It also explores how climate writing nests Anthropocene narratives within the prickly embrace of both a realist approach to climate change and a non-representative viewpoint. As mentioned in Chapter 1, overcoming the limits of representation of unseen subjects requires relinquishing mimetic attempts of traditional nature writing in favour of non-representational emphases on 'the fleeting, viscous, lively, embodied, material, more-than-human, precognitive, non-discursive dimensions of spatially and temporally complex lifeworlds' (Vannini 2015a, p. 318). Put simply, this entails imagination, and specifically in groundwater narratives, subterranean imaginaries. Climate writers who give attention to narrative groundwater reveal the differences between shady reality and shaky attempts at non-representation.

Bodies in Subterranean Places

Immanuel Kant's conjecture that we can only know place through our bodies sits uncomfortably in groundwater narratives. 'The most intimate as well as the most consequential inroad to place is through the body' (Kant quoted in Casey 1997, p. 251). Casey agrees with Kant's proposition: 'To be is to be in place – bodily', or as Casey adds, 'at least bodily' (Casey

1997, p. 399). This assumes an individuated, anthropocentric human body. To write of or in place as a bodily exercise emphasises an immersion in sensory physicality, which arguably brings immediacy to the work. As Mitchell Thomashow (2002) suggests, engaging the senses, the corporeal, is part of writers' attempts to perceive and reveal connections to places. Applied to groundwater narratives the idea of literal human bodily immersion is not immediately evident. A writer cannot, at first glance, 'be in place' in an aquifer. Neimanis stirs up the trouble of the anthropocentric positioning by figuring embodiment as watery (Neimanis 2017, p. 2). Both human and non-human actants, when thought of as bodies of water, are never autonomous, always in relation as they exchange waters throughout lives and existences. Narrative groundwater asks writers to consider the porous and permeable boundaries of bodies, both human and nonhuman bodies, and to consider the task of imagining groundwater, not just remotely, but with porous connections between our shared watery bodies.

To know and write about, or preferably *with* a subject such as groundwater is affected by many porous moments. Iovino connects 'a landscape of porosity' with cognition, saying that '"embodied knowledge" is rooted in the mutual porosity of bodies and the world' (2014, pp. 102–103). She goes on to propose that:

> Knowledge – human and nonhuman informational exchange with the world – is a form of porosity; it is the way the world enters and conditions habits of living [and knowing], thus determining the ways living beings *in-habit* the world. To say that knowledge is 'embodied' means that the world acts together with bodies, becoming sedimented in and filtered through cognitive practices.
>
> Iovino 2014, p. 102

Applying this idea to how climate writers may know of and write *with* the world of aquifers and groundwater flows illustrates not only the shared porosity of our interactive bodies of water, but also how each of our bodies of water, humans and aquifers, influence what we 'know' about the other. Groundwater narratives emerge from watery bodies and minds. As Neimanis explains, 'Just like bodies of water, stories are rarely autochthonous; they usually begin in many places at once, with many unspoken debts (Neimanis 2017, p. 8). Groundwater narratives may be grounded in the subterranean, but they express many voices, many watery places.

Water's mysterious and dangerous movements underground have long attracted cave divers. Limestone caves in karst landscapes offer a place where brave souls may immerse themselves in a form of groundwater. I have walked through limestone caves, had water drip on me from stalactites and stepped over subterranean streams in floodlit caves. However, I am not a cave diver. Robert Macfarlane's (2019) expeditions to underland tunnels, caves and crevices are not for me. Near suffocation, confusing subterranean mazes and gazing blindly into dark, bottomless voids inspired his breath-taking book *Underland* (2019). Bodily entrapment in subterranean water spaces came to global attention in June 2018, when the world watched agape as twelve young Thai boys and their assistant soccer coach were lost, found and finally rescued after being trapped deep underground for ten days. Tropical storms had increased groundwater movements through deep chasms in the Tham Luang cave complex in Thailand. Rising water levels had trapped the footballers on a slip of ground deep in the cave system (Aronson 2019). Global media focused world attention on dangers of a watery underworld. Narrative groundwater spoke in this story as a deadly threat; it engendered terror of subterranean drownings. Groundwater narratives may follow the lead of bodily immersions; the unexpected movements of groundwater drove the narrative as much as the courage of the subterranean cave divers.

Subterranean imaginaries encourage climate writers to consider not only caves, but the pore spaces in rocks. The sandstone that groundwater seeps through, occupying multiple shifting figurations, is not a place I can bodily occupy or observe. Physical senses of sight and hearing are mainly limited to surface atmospherics. Where water fills the porous cavities between rock particles, it is human imaginings of 'geolinguistics' (Le Guin 1984) that may bring such 'places' to human attention. Imagining the water in our cells, in our blood, in our brains creates links to geological porosities. Imaginations of subterranean watery places may need a new style of language.

Stacy Alaimo (2010) advocates strongly for the potency of enmeshments of human and nonhuman bodies. She says, 'Imagining human corporeality as trans-corporeality, in which the human is always intermeshed with the more-than-human world, underlines the extent to which the substance of the human is ultimately inseparable from "the environment"' (Alaimo 2010, p. 2). Porous spaces between many types of corporealities – human bodies, nonhuman animal bodies, more-than-human entities (such as forests, rivers, ice shelves, grasslands, mountain ranges, and the plethora of bodies of water, aquifers), non-living entities (such as stones, plastic bottles, tarmac roads, computer screens) – all join the fray to influence thinking *with* notions of porosity and trans-corporeality. Serenella Iovino (2014) exemplifies this idea perfectly in her description of the expressions of bodily porosity inherently built into the city of Naples, Italy. Iovino, following Stacey Alaimo (2008), links the trans-corporeality of human and nonhuman bodies through a concept of porosity. Iovino talks of the sponge-like architecture and geologies on which Naples and its cultural constructions are built:

> Porosity means here the permeability not only 'between our flesh and the flesh of the world we are of and in', but also the bodies and discursive worlds in which they are located: bodies 'produc[e] culture at the same time that culture produces ... bod[ies]'.
>
> Iovino, 2014, p.103, quoting N. Katherine Haynes

As a porous place, a city-scape body of material places, Naples's reality exemplifies Deleuze's comment that matter 'offered an infinitely porous, spongy and cavernous texture without emptiness, caverns endlessly contained in other caverns; no matter how small, each body contains a world pierced with irregular passages' (Deleuze, quoted in Iovino 2014, p. 101). Thinking *with* porosity of bodily and worldly places contributes to climate writers' repertoire. Groundwater narratives that take readers to subterranean places are influenced by bodily experiences.

Groundwater narratives are also created through a writer's imaginations as swayed by readings, observations and memories. For example, Haruki Murakami's novel *The wind-up bird chronicle* (2003) directly influences scenes in my fictional manuscripts. This novel led me to consider placing my human protagonist Frankie Pankhurst down an unused, dry well in several key scenes. This action became a critical means to bring the feeling of the subterranean and groundwater closer to the protagonist and the reader. The mineshaft/well provides a place where Frankie's dreams and internal thoughts reveal connections to affective memories and elements of groundwater's ubiquitous sensate potency. Murakami's at times fantastical and surreal style and the portrayals of his protagonist, Toru Okado, repeatedly spending time in a neighbourhood well, inspire an approach to writing subterranean imaginaries, albeit an embodied experience of being underground, in the dark. About half-way through *The wind-up bird chronicle,* Toru is stuck for several days in the complete blackness of the well he has been visiting, someone having mysteriously removed his rope ladder. Murakami describes Toru's 'fragmentary memories' arriving in silence, 'like water slowly filling an underground cavern', a somewhat prophetic indicator of later events (Murakami 2003, p. 266). Murakami goes on to describe Toru's annoyance at a seemingly trivial memory turning to anger and then rage that blots out his hunger and his fear at being permanently imprisoned in the well. The first scene where my protagonist, Frankie Pankhurst, spends time in a deserted well/shallow mineshaft, is an unplanned occurrence for her, as she attempts to rescue a dog. With no knowledge that rescue may be coming, her memories and feelings are intensely fearful. The second and subsequent visits to the mineshaft are Frankie's conscious decisions to 'go to the source', to go 'beneath the surface', to become closer to the underground sources of groundwater, as her recently deceased father had regularly suggested she do.

With its history of alluvial gold mining, there are many abandoned wells and mineshafts in the bush around Central Victoria, Australia where the story is set. I once had the experience of rescuing my dog from a similar deep, dank hole. We were taking a morning walk through local bushland, when the dog bolted after a rabbit. I eventually found the dog whining in a vertical shaft about five metres deep, a remnant of a gold miner's hand-dug efforts. I walked to a nearby farm to find assistance, and borrowed a ladder from a helpful old woman, whom I remember was wearing a pink chenille dressing gown. Back at the mineshaft and not thinking of how I would climb out, I was reached my feet down to the ladder and descended into the hole. I spent a few uncomfortable hours stuck in the dark with the dog, because the ladder was too short for me to climb out, with or without the dog. The elderly woman left me and the dog in the gloomy prison to drive some distance to recruit assistance from some nearby campers. They brought ropes and pulled first the dog then me out. I remember the rope burns on my hands and waist. I remember waiting for help, the musty smells, the light above me. I drew on these memories of my trapped feelings, combined with my memories of the intensity induced by Murakami's well scenes, in my descriptions of Frankie's experiences in the mineshaft in the following excerpt. In Murakami's portrayal, Toru Okado's dream-like experiences in the well may be real or illusory. Frankie's reflections, though at times oneiric, are part of her journey towards understanding herself, and her capacity to speak out about groundwater. The blending of what Frankie has learned about the science of groundwater, as had I the author, with my memories of underground experiences, nourished the writing process in creating these scenes.

Extending an uncomfortable bodily experience in the mineshaft to imaginaries of inaccessible subterranean worlds is part of Frankie's decision to literally stay with the dark. These are critical scenes in *Why we cry*, indicating Frankie's journey towards understanding her links to groundwater. The scene is also influenced by Robert Macfarlane's aspiration for his book *Underland*, to be a 'story of journeys into darkness and descents made in search of knowledge' (2019, p. 17). Frankie's physical proximity to sedimented rocks, to cool air, to frogs (her favourite animal) in an underground place initiate her imaginings of groundwater, and hint towards their trans-corporeality, or to use Barad's (2015) term, trans-materiality. My decision to link knowledge about human porosity to groundwater through Frankie's experiences in a deep and silent place involved exploring how to lean towards non-representation (which I discuss later in this chapter), shifting the story away from the broad daylight of everyday events. The mineshaft/well scenes encourage the reader to imagine and feel the proximity of the subterranean, to take the 'inward migration' (Wright 2022), rather than think of it as something simply 'under there'.

Excerpt 5.1: *Why we cry*, Chapter 8

On the weekend, heat blared from an empty sky. Frankie carried her new extension ladder to the mineshaft; or was it a disused well? Nobody really knew who excavated here or why. Water or gold, or both. She unfolded the ladder out to full length, lowered it in. This time she checked that the top of the long ladder protruded from the hole. She climbed down into the cool, dark space. She'd planned this moment. She'd been thinking of the well, not as a vacancy gouged from the skin of the earth, but a place where tens of thousands of years were layered dimly before her eyes, an exposure of sedimentation. She wanted a quiet place to think. She left behind the glare, the heat, the blinding summer light, the crashing pressures of too much to do. She breathed musty air, felt underfoot the pebbled base. The top of the shaft haloed above. She leaned her back on the earthen wall of the shaft, closed her eyes, let the deep force of earth hold her still. She felt her breath cool in her nostrils. Her thoughts trickled, wandered.

Frogs sometimes live in the secret world of wells, where water seeps to fill the space. Moistens their skin. She imagined the nictitating eyelid of a frog, wishing for one extra layer of protection to draw across her eyes. She licked her lips. Her thoughts wandered. Wells are connections to aquifers. Domed brick lids, buckets on ropes. Hand-dug wells, hand-dug mineshafts. Holes have been dug into the earth for thousands of years. Frogs have found these spaces.

Beneath the faint rustle of wind in the trees above, a sensation rose, like cyber-waves, whirr of technology, electrons in metal wires. Beyond vibrations of her damaged tympanic membranes, pulses in her veins vibrated. Too much screen time, she wondered. But different. She tried to see what she heard. Two senses merged. She observed and listened through her skin. Rhythmic thump of a loosened tabla drum, goat skin, thumping, pulsing. Pitched low, profound. She held her palm over her heart. Her whole body absorbed her surrounds. Air, earth, water. Her skin called for water. Porosity of stone, calling. Micro-spaces in her bones thrummed. She was leached, beached, run aground. Memories flashed.

Moments of terror experienced by a frog stranded on concrete on a scorching day, assaulted by wind or dust. A human approaches, mountainous, a hand extends, pink, threatening, like an intergalactic monster. The frog leaps for safety, leaps and leaps to get away.

Frankie's legs twitched. Pain, terror of the dry. She touched her arm, felt her face, held her throat. Skin of the frog world. Alerted to danger through her eyes, her mouth, her tongue, her ears. Her heart banged. Sounds became sensations, came and went. Her thoughts leapt, her body cooled. Frankie let her imaginings ebb and flow.

A chorus of voices murmured, indecipherable utterances. Deep time swallowed sounds, stuttering, whispering. Repeats, cycles, ancients build the chorus. Groundwater's song, rumbled lines on loose kettle-drum skins.

Water plied the surface, followed laws; gravity pulled it to lower places. It juiced the land. Became the inheritance from old rain, mother rain.

Perhaps it fell fifty million years ago. Some dropped only a few thousand years ago; some was pitched to splatter dust in last week's storm. A torrent. Replenishes fissures and micro-spaces at the surface, tops up shallow coffers. Tree roots sing in relief. People dance in mud, singing high trills.

Diviners. Divine people follow water's traces, through deep rift valleys, across deserts, seeking survival. Lennie holding out his sticks. Fossickers. Skin scratchers. Survivors found water, sipped from springs.

Deep, dark, slow, infinitesimally slow. Liquid, fluid-filled, saturated spaces, fractured rock, subterranean stones. Water-filled interstitial spaces, dissolved minerals moving. Soaked salts from rocky substrates, tang, saline suspensions, carried to the surface. Sulphurous farts lingered sour.

Sweet water, drill deeper, deeper for sweeter. Taste the bore, lick the tap. Taste the bath, bitter. Taste your clothes. Mineral salts. Earthy. Borers' sweat mingles with salty flow. Taste the bath. Taste your clothes. Taste the bath, suck your sleeve. Earthy mineral salts with moving electrons, moving energy, the start to life.

Frankie licked her arm. Closed her eyes again to listen.

She remembered stories of cave divers. Images from tourist brochures and news clips flashed. People, turtled-back with scuba tanks, diving in deep-pooled caves, exploring pearled depths, turbulent tunnels, looking for children. Crablike in crevices, bubbling through faults, through vaults into darkened caverns. People swam in endless embrace, floated on ageless flows. Some were lost in the labyrinths and drowned, gone too far, lost the safety line. Frankie shuddered.

She shook her head. The double life of amphibians knocked, asking to be let in. Am I frog on land, on water? Am I tadpole, becoming frog? If not, becoming what then?

Frankie felt an unknowable cavity yawn inside. Up there, water cycles lurched out of shape. Becoming something, always becoming. Unknowability of self, the indeterminacy of aquifers, same same. Her sense of self felt as intangible, as unreachable as subterranean flows. Yet so surely there, living, moving, becoming. Quickening.

Frankie stayed in the mineshaft for an hour or so, breathing, thinking, dreaming. Water seeped below her on an underground sojourn, soaking through capillaries, seeping into fractures beneath her cavernous refuge. Moving, ever deeper through the earth. Seeking release. Seeking release. A beetle scurried, a moth batted its wings in the dark. Her legs ached, she stood to stretch, her hands searched the darkened space.

She climbed back to the surface, blinked in the glare, warm air on her skin. She lifted the ladder and carried it back to her shed. She descended into the mineshaft again the next day, and the day after, and continued for a week. Each descent brought new waves of fear, of calm, of numbed loss of senses, of solace. She was at times seduced by the dark, captive, held in the moments of seeing and not seeing the tangible earth around her.

She felt at times looked upon, as if the walls of the shaft had vision. She felt porous. Felt the permeability of the earth around her. Gravitational heaviness and the lightness of angels passed, inside and out. Each visit to what she secretly called her tabernacle gave respite from surface distractions, brought her closer to imagining a watery life, an amphibian life. Each layer of sedimentation spoke through its exposed surfaces on her outstretched hand. She inhaled dank smells of water. She entered a time where simple statements of truth jumped into stories, a time where water flowed.

This was not a fall.

'I have water in my bones.' Her voice croaked.

Water circled through rock, roiled, coiled, drawn by gravity, timeless, slow. Invisible. Calling. Subterranean water, flowing through geology's underworld. It moved tentatively, as if in search of relief. Crevices diverted the flow. Waters silently passing time, moving through warped spaces, indeterminate. Water filling air spaces, lubricating life between grains of rock, soaking, soaking, seeping. All that is solid, all that is fluid. Continuous flow, incremental creep. Deep, slow seep. Deep seep. Pinholes bleed, water retreats, withdraws, beyond reach. At the surface springs, swamps, leakages from stream beds. Grounded no more.

Groundwater narratives are less concerned with trying to know the physical boundaries of an aquifer's reach and more interested in expressing groundwater's relational capacities through storying. It is not only through science, nor only through being down in the mineshaft that Frankie found meaning in groundwater. Through her interactions with and memories of her father Lennie, the water diviner, through her learning with Heather the hydrogeologist, and through conversations with her Aboriginal friend Wade, groundwater's presence gains potency. Frankie's 'inward migration' (Wright 2022) and her relational, imaginary encounters with groundwater aim to expose readers to narrative groundwater's potency.

Place-Based and New Nature Writing

Consideration of place and time within the junctures of philosophy, literature and science is not a totally new endeavour. Thinkers, scholars and writers have for centuries sought to locate human-made meaning within the elements of place and time. 'Where' and 'when' are locators of meaning and important elements of plot in climate writing, particularly fiction. Explorations of the influences of various places on humans have shifted, with greater attention more recently given to the influences that humans have on place.

Place-based literature and nature writing has historically advocated connections between humans and the 'natural' world. The term pastoral describes early forms of nature writing. The foundations of early rurally based poetry and literature were laid in with Virgil's *Eclogues*, written around 37 BCE. In the United Kingdom and Europe, variations in pastoral writings occurred over centuries, depicting love of serene beauty in the natural world, through to the threats and hardships resulting from appropriation and enclosure of the commons throughout the late 1500s (Williams 1975, pp. 26–28). Romantic poets of the late eighteenth and early nineteenth centuries, such as Elizabeth Barrett Browning, Samuel Taylor Coleridge, William Wordsworth and John Keats, tended to glorify the tranquillity of country life through gentrified observations that left behind the earlier Virgilian depictions of working countrymen and women, particularly shepherds. Themes of love and loss in country life persisted. Seasonality and harmonious peace in the observed natural world prevailed through the American nature classics, which include classics such as Ralph Waldo Emerson's transcendentalist essay 'Nature' (1836) and Henry David Thoreau's *Walden* (1854/1993). Ecocritic Lawrence Buell (2022) analyses Thoreau's place in American nature writing and the contributions that Thoreau made to the large body of literary works that are based on 'environmental imagination'. *Walden* contributed to American naturism, but went beyond the early national literature which focused on mapping and celebrating the land, usually from the colonial purview, assuming without question colonists' occupation of lands (Buell 2022, p. 14). Buell's (2022) examination of the evolution of American nature writing shows how early Western nature writing focused attention on 'man's' connection to, and the increasing disconnection from, the natural world during the nineteenth century and early twentieth century. While at first women were rarely shown entering the wilds, Buell's closer examination of the historical record 'shows that women literary naturists were a strong presence almost from the dawn of American literary emergence' (2022, p. 27). Here is not the place to expound a history of nature writing, it is well traversed by Buell (2022). The focus of this chapter starts with a modern take on the pastoral, or what is considered 'natural'.

Traditional nature writing, even with its problem of the external subject, is not only about nature. It is also about the human experience of nature – it is about human perceptions of our place in the environment. Nature writers traditionally blithely accepted the assumed primacy of humankind. At the same time, paradoxically, nature writers often remain deeply concerned with expressing their own interiority, their own 'fates' in connection to the places they enter (Tredinnick 2005, p. 25). As Timothy Morton (2007, p. 125) argues, in traditional nature writing, by attempting to describe 'nature' as something separate to humans, as something 'over there', human and nonhuman separation from pristine places is accentuated rather than abolished. The concept 'nature' is critiqued as being nonexistent, it is rarely pristine (Soper 1995; Morton 2007). Nature, however, has never been silent about human violence at its expense. Nature has always 'spoken', made its representations to humans (Carter 2015). What we 'hear' and 'see' is becoming more and more important in the context of global warming. A nature writer might think of groundwater as a place – consider its emergence in a spring, or its flows in the confines of an aquifer. But as a deep subterranean entity moving, acting, speaking its own language, and ever becoming something else, groundwater, as a posthuman material entity, becomes no longer place-like, no longer fixed, no longer separated from humanity's foibles.

More recently, in what is termed *new nature writing*, writers challenge preconceptions of nature as having limited agency, and shift cultural biases that have prevailed as they 'cordoned off scenery into pretty ghettos' (Buell 2022, p. 4). Contemporary new nature writing considers many things 'from nature's point of view – politics, aesthetics, language, authorship, morality, culture, humanity … It takes an ecological view of all things' (Tredinnick 2005, p. 25). It draws attention to anthropogenic 'mistakes' of representation, aiming to change preconceptions, cultural biases and mindsets (Tree 2020, p. 11). One way that writers might imagine the complexities of ecological connection is to accept that nature 'over there' does not exist, Morton (2007) argues. Against the 'false immediacy of ecomimesis' Morton (2007, p. 187) suggests that an aesthetic dimension, an ecologically political *noir* form, can undermine mimetic stories about being involved with natural places. The tones of groundwater avoid mimetic expressions. As is explored later in the chapter, Nigel Thrift (2008), and more vigorously Phillip Vannini (2015a), describes the impossibility of accurately representing the diversity of lifeworlds. Rather than writing sweet stories of beautiful souls' connections to nature, Morton suggests that the melancholy of *noir* forms of narration, where the shadows of ecological catastrophe are preserved, can be part of a posthuman writing that enmeshes human responsibility with the catastrophes in which humans are implicated. He adds: 'Instead of whistling in the dark, insisting that we are part of Gaia, why not stay with the darkness?' (Morton 2007, p. 187). This aligns with Lawrence Buell's (1998) suggestion of the importance of acknowledging 'toxic discourse', of disaggregating the pastoral idyllic, to engage with environmental justice literatures that address the increasing toxification of environments. Writing the threats to groundwater reserves through pollution or over-extraction through the lens of toxicity or melancholy *noir* not only moves groundwater away from idealised place-based nature writing, but increases the chances of giving groundwater an agential voice, enabling groundwater to interact with multiple actants of its subterranean realm and surface impacts in a quest for environmental justice.

Buell analyses Rachel Carson's classic non-fiction *Silent spring* (1962) as a contemporary initiator to literature that may be called toxic discourse. Based on her indictment of DDT and agricultural herbicides and pesticides, Carson's meticulous research and lyric writing is credited with not only revealing the dangers of chemicals to human health, but to opening community and political awareness of human connections to their environments. A risk of Carson's work however, argues Buell, is that it continues to mythologise a generalised, utopian pastoral, where Carson suggests that the damage done by dangerous chemicals divorces whole populations of people from possible idyllic relationships with the environments they inhabit. Rather than longing for pastoral havens, Buell (1998, p. 665) argues that reading *Silent spring* as toxic discourse enhances environmental imaginations that put communities in relationships to environments as collectivities rather than as individual consumers or escapees. Toxic discourse links affected populations to an environmental justice movement that seeks redress and mitigation of toxicities (Buell 1998, pp. 647–648).

Critics of representational place-based writing such as Val Plumwood (2008), Lawrence Buell (2001, 2022), Ursula Heise (2008) and Adam Trexler (2015) have argued that writing to celebrate place cannot alone answer global environmental concerns. Plumwood argues that 'We need to replace loose discourse about "sense of place", … by place-based critique, that can make room for the power analysis of an environmental justice perspective' (2008, p. 140). Maintaining realistic expressions of groundwater's influences and its plight due to excessive extraction, alongside or braided with speculative imaginaries, is one way to maintain politics and principles of environmental justice in groundwater narratives. In his critique of representational nature writing Buell explains:

> nature has been doubly otherized in modern thought. The natural environment as empirical reality has been made to subserve human interests, and one of these interests has been to make it serve as a symbolic reinforcement of the subservience of disempowered groups: nonwhites, women, and children.
>
> Buell 2022, p. 21

Writing groundwater narratives from an empowerment model means mounting a critical stance on the assumed disempowerment of 'others' and 'other places'.

Privileging concepts of ownership, or systems of commodity exchange that lead to exploitation and fragmentation of ecologically linked places, disregards ecological enmeshments. Think monocultural, industrial-scaled agriculture, or for example think how groundwater, or forests, or coal and oil reefs are treated only as resources to be mined, that rivers are places to be dammed. Foregrounding principles of environmental justice reveals the direct responsibility of colonial governments, Western political leadership and pervasive materialism for creating and now seeking to maintain catastrophic climate-changing activities (Ghosh 2021). Writing against such forces is the challenge of climate writing. Environmental justice perspectives in writing *with* places include an analysis of colonial and corporate power, enabling otherwise unheard voices of resistance to be expressed.

A barely told groundwater narrative about the impacts of aquifer pollution from per- and polyfluoroalkyl chemical substances (PFAS) around airports illustrates the literal and figurative porosity of storying. Porous spaces in sandy soils and subsoils carry pollutants. PFAS are a large group of human-made chemicals found in many products including cleaning products, some textiles, cured leather, paper and paints, fire-fighting foams and wire insulation. Australian firefighters have for decades used fire retardants containing PFAS in training. Illnesses that are known to be connected to PFAS exposure (most commonly occurring in polluted drinking water and Teflon pans, as well as fire retardants) include non-Hodgkin's lymphoma and kidney, testicular, prostate, breast, liver and ovarian cancers, as well as liver damage, decreased fertility, and increased risk of asthma and thyroid disease. The Williamstown, NSW RAAF airbase, adjacent to Newcastle Airport, is one example among many where PFAS chemicals have seeped into aquifers. The *Four corners* program presented on ABC (Australia) television by journalist Linton Besser on 9 October 2017 revealed the wide-ranging impacts from the toxic groundwater pollution from fire retardants used around Australian defence force airports. The emotions of local landholders affected by the pollutants ranged from rage to despair (*Contamination* 2017). Redress for property owners has been recognized in Australia, but, at the time of writing, affected residents around the country continue to join a class action claiming that current compensation offered is insufficient for damages to health and the loss of property. Groundwater performs its far-reaching impacts in these current, real-life stories, where environmental and social justice principles overlap. Interrelationships between porous subterranean places and the effects of groundwater pollution from PFAS are ripe materials for writers of groundwater narratives.

Nature writers often anthropomorphise expressions of place in the more-than-human world. The charge of anthropomorphism, meaning to present nonhuman worlds as agential, intentional or illegitimately displaying human characteristics, has been a mechanism to police and maintain

Cartesian, dualist beliefs that humans and nonhumans are inalienably different. According to this outmoded idea, humans have minds and sentience, a capacity to love, and nonhumans do not. Eco philosopher Val Plumwood urges writers to:

> Free up your mind, and make your own contributions to the project of disrupting reductionism and mechanism. Help us re-imagine the world in richer terms that will allow us to find ourselves in dialogue with and limited by other species' needs, other kinds of minds.
>
> Plumwood 2009, p. 127

Attributing human-like responses to groundwater's expressions flouts the sin of anthropomorphism, as Val Plumwood suggests. By 'sinning' writers aim to develop an ecological sensibility that connects humans beyond our usual limitations (Plumwood 2009). Adopting the human lens of creativity, writing groundwater narratives becomes a 'heuristic strategy' (Iovino and Oppermann 2014, p. 8), which aims to reduce the ethical and textual distance between human and nonhuman characters. Anthropomorphic options become feasible. For example, ascribing a metaphor of female strength with the mountain, Ursula Le Guin's (2006) essay 'A very warm mountain' describes the Mount St Helens volcano blasts in 1980, at first using reports recounted by print and radio media journalists to describe the eruption. She interweaves 'factual' descriptions of mud balls, ash and darkness with her own experiences at the time. She ends by evoking the volcanic blasting of Mount St Helens in terms of her own characteristics as a fiery woman. The mountain changes from the grace and stability of being dressed as a woman, with cloud-like hat and scarf, and breast-like in form, to something out of control. The exploding mountain expresses Le Guin's own aspirations. She refuses to describe the mountain's rapid change as a desiccating cancerous form, blown to pieces, preferring to see the volcano as a raging, farting, screaming woman, alive and healthy still. To give gendered characteristics to groundwater is tempting; she flows, she fills the spaces of her domain, she brings life to the surface, she is sucked dry by extractive industries. Ascribing female characteristics to groundwater is a way to politicise its presence.

Paradoxically, being prepared to embrace anthropomorphism rubs against efforts to push anthropocentrism aside. Climate writers seek to unsettle the privileging of human-centred perspectives as they give voice to nonhuman and more-than-human entities. This may at times involve anthropomorphic expressions that elaborate how human characteristics can be seen in the actions of nonhuman animals and more-than-human entities. Maintaining a critical awareness of anthropocentric tendencies in climate writing remains essential (Clarke 2015; Johns-Putra 2018). There is a certain hubris in assuming that human features belong only within the human species.

Realistic Imaginaries of Place in Fiction

The material realities of the impacts of global warming, alongside the reality of receding and polluted aquifers, colours what a political and ethical literary response may be. In realist-styled fiction, climate writers aim towards shining a light on what would otherwise not be considered (Dillard 1982; Boulter 2007; Wood 2008). For example, creating stories of unsustainable groundwater management responses by water authorities and water traders may illuminate otherwise untold narratives. Writing groundwater narratives entails being wary of attaching a 'cosy' view of groundwater's singularity. There is no one aquifer, no one author. Decisions to express multiple voices of a more-than-human entity, groundwater, in the Anthropocene enable mutual performances between authors and aquifers, acknowledging both as agential subjects.

In the history of the novel, as far back as the Greek novels analysed by Mikhail Bakhtin, there are no 'pure forms' (Bakhtin 1981). As many literary critiques, including Watt (1957), Bakhtin (1981), Morris (2003), Mulhall (2009) and Vermeulen (2015), point out, the history of the literary novel does not unfold neatly. Rather, the various conventions for conveying meaningful impressions of human and nonhuman interactions in fictional narratives have overlapped and interjected on each other for centuries. Female writers, such as Aphra Behn (1640–1689) and Lady Mary Wroth (1586–1652), writing in a realist style in the seventeenth century, were the first to adopt a style of novel writing that blurred fact and fiction. They are considered 'mothers of the novel', despite the conventional turn to Cervantes writing in the early 1600s, or Defoe and Fielding writing in the eighteenth century, as the early 'fathers' of the novel (Spender 1986). Realist fiction's traditions are rooted in Enlightenment thinking that stemmed from the burgeoning trust in the objective scientific knowledges of the eighteenth and nineteenth centuries. Critiques abound from feminists, modernists and postmodernists of those daring to assume fidelity to not-knowable truths (Morris 2003; Clarke 2015; Johns-Putra 2018). Assuming universality in patriarchal discourses in science and society is shunned in feminist critiques (Irigaray 1985, 2002). Groundwater narratives in the Anthropocene become part of this porous and moving trajectory of storytelling through various forms and styles of the novel.

Definitions of literary realism are almost impossible to pin down (Morris 2003). Traditionally, realism purports to produce an 'authentic account of the actual experiences of individuals' (Watt 1957). Defining literary realist style as 'writing that is based upon an implicit or explicit assumption that it is possible to communicate about a non-textual reality beyond the writing' is a starting point (Morris 2003, p. 6). I make the distinction between historical realism in fiction and a realistically styled form of storying. A realistic style of writing illuminates, rather than copies, reality (Felski 2008). Despite the criticism that attempts at verisimilitude in transcriptions of actuality are not possible, and do not necessarily produce a work of truth, a realistic style has persisted as a broad approach to novel writing (Morris 2003, pp. 4–6). Humans, it appears, relate to realistic stories.

I do not here enter the already well-canvassed debates about what constitutes the various merits of realist, modernist and postmodernist literature in relation to climate writing's address to the Anthropocene. A very brief outline of features of each of these styles follows. Realist-styled fiction addresses comprehensible subject matter, builds a cohesive (perhaps at times variable) point of view and a credible chronology and ordering of events across plausible timeframes (Morris 2003). In comparison, modernist writers such as Virginia Woolf, James Joyce and T.S. Eliot warp chronology and reflexively acknowledge the fiction of novel writing and poetry. They invite readers to join them and their characters in questioning the worlds they compose (Morris 2003). Postmodernism goes further to question whether objective reality exists at all. Postmodern writing breaks chronology, disrupts point of view, fragments texts and is totally self-reflexive. Adeline Johns-Putra makes a convincing case for strategies of postmodernism, particularly that of metafiction, which I discuss below, as a way to decentre or elide human agency in climate writing (Johns-Putra 2018, p. 29). Creating a fictional world that draws upon subterranean imaginaries enables authors to create narrators who question themselves and their relationship with groundwater utilising any of these styles.

Barbara Kingsolver's best-selling novel *Flight behaviour* (2012) is an example of a work of realist fiction that addresses impacts of climate change. Kingsolver's protagonist, Dellarobia, has a spiritual-like experience in discovering the massed monarch butterfly colony in a pine forest on her family property in Tennessee. She questions the scientists, activists and naturalists coming to observe, and in their different ways attempting to protect, the stranded colony of monarch butterflies. The rare butterflies are at risk of habitat loss because of her family's plan for deforestation of the mountainside where the butterflies have nested. Like Kingsolver in *Flight behaviour* (2012), climate writers aim to complicate the stereotypical portrayals of scientists, activists and climate deniers in order to emphasise diversity of perspectives and characterisation, and to enrich the breadth of possible relationships between diverse characters and groundwater.

At the end of *Flight behaviour*, Dellarobia decides to study science as a way to engage in the climate change dilemmas she has discovered. Science and scientists are placed as 'hero' figures. The 'objectivity' of science, while portending to be value-free, is presented as having high moral values attached (Sleigh 2011, p. 127), particularly under the pretension of lepidopterist Ovid Brown's humility. In Kingsolver's novel 'the effects of climate change are disproportionately borne by those who cannot afford them' (Trexler 2015, p. 228), asking the reader to consider class issues in climate change responses. Kingsolver reinforces the potency of a contemporary, everyday story by giving voice to a protagonist, Dellarobia, who learns of the complexities of sciences and environmental activism, and deals with deniers of climate change's impacts. Set in Appalachia, a characteristically southern American region, Kingsolver makes little mention of First Nations perspectives on the questions she explores, a significant gap in the narrative.

As modernists and more recently postmodernists have proposed, metafiction describes fictional writing that 'self-consciously and systematically draws attention to itself as an artefact in order to pose questions about the relationship between fiction and reality' (Waugh 1984, p. 2).[1] Readers' attention is drawn to their own interpretations of the text they are reading (Johns-Putra 2018, pp. 29–30). J.M. Coetzee's novel *Elizabeth Costello* is exemplary of this self-conscious form of fiction, which, within its own terms, questions the 'reality' it describes (Coetzee 2004; Mulhall 2009). In metafiction the narrator is strongly present, with all their self-proclaimed limits, and the reader is refused the role as passive consumer. The narrator invites the reader into her/his own questions. Self-consciously creating a fictional world with human and nonhuman characters embedded in places invites readers to question the narrator's and/or the author's imaginary intent (Waugh 1984, p. 12). Coetzee's narrator, Elizabeth Costello, has the reader questioning reading, writing and philosophy of human and nonhuman relations. As Costello prepares her thoughts for a conference paper she is about to deliver, she reflects. 'Specifically, she is no longer sure that people are always improved by what they read. Furthermore, she is not sure that writers who venture into the darker territories of the soul always return unscathed' (Coetzee 2004, p. 160). Metafictional techniques include creating self-reflective narrators, who question the questions the author is exploring. A human character's thoughts and observations, as well as groundwater's movements and responses, give expression to groundwater's voices. Through metafictional devices, groundwater's textual performances are in dialogic relation to its own ecological reality. Elements of metafiction can support reflections on narrative groundwater's efforts to be 'placed' in its many localities. Drawing on subterranean imaginaries, writers test inventiveness in language at the same time as self-consciously questioning various realistic expressions of unseen, timeless watery matter. Groundwater narratives question possibilities of strictly 'true' or realist representations of a purportedly invisible subject.

Subterranean imaginaries speculate how the unseen might enter the world of human and nonhuman characters. Subterranean imaginaries call for portrayals of the interiority and partiality of the fictional world of human and nonhuman protagonists. At the same time, in embracing partiality and inconclusiveness, Bakhtin reminds us that, 'since the novel is aware of the impossibility of full meaning, presence, it is free to exploit such a lack to its own hybridising purposes' (1981, p. xxxiii). Hybridity is characterised by ambiguity. The writer of groundwater narratives signals the inability of texts to convey conclusive, fixed meaning. Hybridity in groundwater narratives is another way that climate writers shoulder their task to embed and embody humans and more-than-humans in closely knit circumstances.

Science fiction, speculative fiction and dystopian fiction are other forms of storying with the elegiac, invisible qualities of groundwater. Dystopian speculations have been the main form of climate writing, particularly in the recent wave of climate fiction that draws attention to cataclysmic futures (Clarke 2015; Trexler 2015). Considering the invisibility and limited knowability of underground water, dystopian speculations may appear inevitable. Given the harsh reality of anthropogenic threats to groundwater sustainability both in Australia and worldwide, whether utopian stories of hope or dystopian speculations do justice to the real urgency is perhaps a fruitless consideration. Michael Segal worries that, in the public realm, both dystopian and utopian narratives 'have a tendency to be useless. They leave the public spectating a stalled debate between extremes and generate ample motivation to check out' (2017, p. 126). Segal weighs in on the importance of verifiable science in climate narratives. As mentioned in Chapter 3, the problems of 'dystopia creep' also press against the paralysing potential of dystopian worldmaking that looks increasingly like the contemporary world (Mills 2022). Groundwater narratives aim to express groundwater's potency, its complexity, its vulnerability in ways that might influence readers' considerations of not just their knowledge about groundwater, but also their capacities for responses, actions and agency towards vulnerable environmental entities. Expanding writing approaches to storying *with* groundwater through subterranean imaginaries asks for braiding of both imaginary and realistic elements in the narrative so as to ensure political relevance and a capacity to influence change. This is discussed in more detail in Chapter 7.

Writing Indigenous Places

First Nations scholars and writers are increasingly writing of the rupture to their connections to place. The power and the vulnerability of First Nations' interrelational connections to lands and waters, particularly following the bodily consequences of colonisation, massacres, murders, and theft of land, language and culture, have prevailed despite the worldwide colonial project (Ghosh 2021). Writing of place in an Australian context, and in many colonised countries, is imbued with stories of tragedy, resistance and resilience. Alexis Wright's (2022) missive holds firm, particularly for First Nations writers: take the solitary 'inward migration', preserve oneself 'locked in the prison of the mind', retreat to 'the dwelling place of stories, a return to Country, going home to where the stories of our culture are kept in the mind' (Wright 2022, p. 58). In Australia, the term Country unambiguously includes references to Indigenous peoples' belonging, inhabitation and spiritual relationship with the territory of their clan, the places they were born and places that 'grow you' (Neidjie 1986, 1989; Bowman 2015). Sovereign custodians of colonised lands spurn the 'white possessive', in preference for culturally embedded familial relations and responsibilities for the trees, rivers, mountains and forests of their terrain (Moreton-Robinson 2015). Indigenous resistance to settler understandings of Australian landscapes have often been ignored, never so seriously as through the original mythologising of Australia as an uninhabited land in the term *terra nullius*.

From an Indigenous perspective of Country, a silent landscape is implausible. Much Indigenous cultural lore and ancestral beliefs are strongly linked with land and water (Strang 2013, p. 189). Examples of Indigenous voices entering the Australian nature writing domain include the collection of stories *Every hill got a story: we grew up in Country*, compiled by the Central Land Council, based in Alice Springs (Bowman 2015). *Every hill got a story* recounts the lives and connections to Country of numerous Central Australian Indigenous families, illustrating the depth of spirituality, sustainability and the power of storytelling. Similarly, several essays and the book *Songspirals* (Gay'wu Group of Women 2019), written on Bawaka Country in Arnhem Land in northern Australia by collaborations between Country's more-than-human voices, Indigenous voices and white settler voices, represent long-term engagement aiming to contribute to responsive relationships between people and places (Gay'wu Group of Women 2019; Wright *et al.* 2012; Bawaka Country *et al.* 2016, 2018, 2022). Founded on Indigenous epistemologies, the 'we/us/our' in these works are handled with great care and fluidity, so as to include Bawaka Country as lead author. The writing acknowledges 'the diverse land, water, human and nonhuman animals, plants, rocks, thoughts and songs that make up Yolŋu homeland of Bawaka in North East Arnhem Land', particularly referencing the more-than-human entities and kin that are 'encompassing, alive and agentic' (Bawaka Country *et al.* 2022, p. 436). The voices of Country are never quiet in this spiralling of collective voices that exercise connections and responsibilities.

Australian Indigenous people's connections to cultural waters are significant in understanding and writing about stewardship of place and waters (Langton 2006; Moggridge 2010; Marshall 2017). As Marcia Langton discusses: 'In the cosmologies of Australian Aboriginal peoples, water is a sacred and elemental source and symbol of life' (2006, p. 139). Traditional cognitive, social and temporal practices of water management, as explained by Elders of specific communities, may often be indiscernible to the white, scientific gaze (Langton 2006, p. 141). Langton's work introduces ethnographic records of Aboriginal accounts of their long-held cultural engagements with waterscapes. The painting *Yarrkalpa (Hunting Ground)* in the National Museum of Australia is a collaboration between eight female artists showing practical aspects of occupying Country, such as burning, gathering plants for food and medicine, tracking reptiles and following waterways. The painting also represents the story of the Seven Sisters pursued by Yurla across the landscape (Mahood 2022, p. 66). Mahood's work with Martu people of northern Western Australia interprets paintings and maps for Western comprehension. She travels Country with local people and discovers the convergence of subterranean water flows to waterholes, revealing Martu knowledge of both subterranean and surface water ways (2022, p. 66). Ageless First Nations' knowledge of subterranean waters in the deserts of central and northern Australia have preserved life and culture in these remote places.

Aboriginal connections to cultural waters are brilliantly expressed in Alexis Wright's two significant novels *Carpentaria* (2006) and *The swan book* (2013), each expressing the significance of water and Country to all their inhabitants. In her award-winning novel *Carpentaria* (Wright 2006), Alexis Wright's protagonists Norm Phantom, Angel Day and their son Will Phantom articulate disdain for white-fella ignorance of Country. Wright's characters embody intimate ancient knowledge of Country and waters. In the opening, dreamlike scenes, aquifers make a rare appearance. An ancestral creative serpent scours rivers and the underground beneath mudflats:

> When it finished creating the many rivers in its wake, it created one last river, no larger or smaller than the others, a river which offers no apologies for its discontent with people who do not know it. This is where the giant serpent continues to live deep down under the ground in a vast network of limestone aquifers. They say its being is porous; it permeates everything. It is all around in the atmosphere and is attached to the lives of the river people like skin.
>
> Wright 2006, p. 2

Later in Wright's novel, the agency of Country is well portrayed in a dramatic scene. Wright gives agency to a killer rock. Will Phantom and his friends set alight a mining explosives site so as to destroy underground storage tanks of fuel, aiming to drive the mining company from Indigenous ancestral territory. The boys run for their lives from the explosion and from pursuit by armed security guards, one being yellow-haired.

> Fate and precious moments are tied up together, and as the saying goes, *What goes around comes around*: the yellow-haired man tripped. Instantly his head was split open at the temple by a rock that had, up to that moment, lain on the ground, embedded in soil that was thousands of seasons old, untouched by humankind since the ancestor had placed it in this spot, as if it had planned to do this incredible thing.
>
> Wright 2006, p. 405

Will Phantom had seen that

> the rock was ready, waiting for the moment ... He had even thrown himself towards the man to try to break his fall. It was too late, a snap, how quickly a driven man could be defeated. Will had no idea a rock could rob him of his revenge ... and there was the stone, still there, unmoved.
>
> Wright 2006, p. 406

Wright has given voice and action to a specific rock in a place, exercising her authority as a Yolŋu writer who consulted with Elders from her Country to write this story.

Another example of talking stones is the Kalapuya Talking Stones near Eugene, Western Oregon, USA.[2] Kalapuya people invite a reclaiming of language and connection to place through these fifteen inscribed rocks throughout the Whillamut Natural Area, near the Willamette River. Visitors read the inscriptions and are invited into the spirit of place. Indigenous peoples worldwide maintain sacred connections to place through both tangible and ephemeral practices: story, song, dance, paintings and written expressions of spirits of the land.

Examining the works of established Indigenous fiction and non-fiction authors is an important way for Indigenous and non-Indigenous climate writers to acknowledge the potency of Indigenous expressions of place. I here make suggestions from Australian First Nations writers and scholars whose voices create interconnections with places. Ellen Van Neerven's short story collection *Heat and light* (2014) is another example of writing that particularly embodies Australian Indigenous connections to place and waters.

Among many Australian Aboriginal writers, the work of Tony Birch (2015, 2016, 2020); Melissa Lucashenko (2013); Larissa Behrendt (2021); Kim Scott (1993, 2011, 2017); and Alexis Wright (2006, 2013, 2016) provide examples of fiction and non-fiction that illustrates Australian First Nations' perceptions of place. Indigenous hydrogeologist Bradley Moggridge's (2010) non-fiction and essays inspire an understanding of Indigenous connections to groundwater. Likewise, Indigenous water scholar Virginia Marshall's text *Overturning aqua nullius: securing Aboriginal water rights* (2017) provides essential information and approaches to the legal and cultural struggle to overturn the myth of *aqua nullius*. She argues that Indigenous peoples' diverse connections to water bodies are embedded in cultural lore, but are largely ignored in Western water management policy. Bruce Pascoe's *Dark emu* (2014) traces Australian Indigenous people's historical connections to place through farming practices. These and many more Australian Indigenous writers illuminate Indigenous knowledges of Country and Waters.

The history of white Australian nature writing is not well known, but has a similar, predominantly racist and masculine trajectory to its American and UK counterparts (Griffiths 2018; Buell 2022). Much nature writing of Indigenous places by white settler Australians remains problematic as the tragic theft of and damage to Aboriginal and Torres Straight Islanders' lands and culture remains predominantly unacknowledged (Rose 2005; Griffiths 2018). The Australian colonial vision of landscape as backdrop to pioneering activities is reflected in much historic Australian nature writing. For example Donald McDonald's children's book *The bush boy's book* (1911) and the ensuing series which promotes settler bushcrafts exemplify masculinised early Australian nature writing, where the skills for survival in a hostile landscape are emphasised. Elyne Mitchell, famed for the children's book *The silver brumby* (1958) and its sequels, also wrote two adult nature books: *Speak to the Earth* (1945) and *Soil and civilisation* (1946). Both these books argue for a colonial land ethic with eloquent nostalgia, seeking to build white settlers' pioneering spirit as they attempted to conquer new and alien geographies. Both books make the long Indigenous custodianship of the Upper Murray Country, where Mitchell wrote, invisible. Blindness to Indigenous presence and relationships to place is a gaping hole in most colonial-styled nature writing.

Contemporary white climate writers who acknowledge and respect First Nations peoples' eons-old connections to place continue to traverse difficult terrain. Fictional expressions of 'ecological imagination' are bound to be 'contingent and uneasy' (Tredinnick 2005, p. 8), because white settler status and inherent, unacknowledged racism binds many writers inadvertently to colonial attitudes. Concealing responsibility for who stole this land, for who refused the use of Indigenous language, and for who took children and adult custodians from communities and Country is common. In ongoing violent contempt, white laws continue to remove Indigenous children from families, continue to incarcerate Indigenous people at at least double the rates of white incarceration. Indigenous health, education and employment indicators trail drastically behind white and immigrant measures. Ongoing racist and colonial-based actions continue in Australia, as evidenced by the 'Intervention' in Indigenous communities between 2007 and 2012, called by the then Howard government the Northern Territory Emergency Response. The ineffective 'Close the Gap' programs on health, education and employment outcomes continue to fail Indigenous Australians. The initial dismissal of the Uluru Statement from the Heart in 2019 by Prime Minister Turnbull's Liberal government was a dismal and cruel response to the ongoing calls for Indigenous rights and a voice in the Australian Parliament. Only recently has the Albanese government committed to a referendum to change the Australian constitution to act upon the aspirations of the Uluru Statement from the Heart. White writers' appropriation of Indigenous stories through the use of Indigenous words and stories continues to take away Aboriginal ownership of their own sovereignty and resistance (Wright 2016, pp. 60–61). An ongoing 'storytelling war', where dominant settler stories and media coverage fail to recognise Indigenous experience accurately, continues to contribute to the silencing of Indigenous stories (Wright 2016). In defiance of such ongoing coloniality, as Zoe Todd says, 'the work of Indigenous scholars and artists promises to speak back, reshape, and change the direction of current human-centric and Eurocentric framings of the Anthropocene' (2015, p. 249). This work will drive the innovation and effectiveness of relationality in groundwater narratives.

Settler writers who express Indigenous viewpoints in stories invite justifiable critique from those who contend that white-fellas, women and men, have no place in expressing Indigeneity in stories. Alternatively, many Indigenous and non-Indigenous writers take the line that it is a greater sin of omission for white writers not to express First Nations' experiences, however partial and limited. Larissa Behrendt's book *Achieving social justice: Indigenous rights and Australia's future* (2003) outlines the intersections of colonial history and law, systematically showing the trajectory of omissions and rejections of Indigenous rights in white mythology, stories and constitutional law. Esteemed Indigenous scholar Megan Davis highlights Behrendt's view 'that space be made within the Australian story and Australian mythology for Indigenous people. White Australia must learn about Indigenous Australia' (Davis 2003, p. 816). Importantly this learning for both Indigenous and white writers comes with and sits alongside forming relationships with the diversity of Indigenous Australia through literatures, and through listening and respectful communications with Indigenous communities. First Nations' culture is not a homogenous body of knowledge. Simply put, white settler writers should never assume to write an Indigenous voice, because of all the problems of appropriation, misinformation and stereotyping that ensue. Third-person observations of Indigenous characters by white characters/narrators also require scrutiny to avoid stereotyping or misinformation. Indigenous writers are able to acknowledge their tribal connections to their specific Country and stories. Different Country tells different stories. Through collaboration, meaningful relationships and decolonising research methods, aspects of Indigenous experience may be expressed by both Indigenous and white writers (De Santolo 2018). White writers are asked to acknowledge their partiality of perspective in relation to Indigenous places or characters. In decolonising storytelling a transformative approach that avoids extraction and exploitation of Indigenous culture and story is called for (De Santolo 2018, p. 206). Climate writers seeking to explore the potency of narrative groundwater start with reading and listening sensitively to portrayals of Indigenous connections to Country and water. For narrative groundwater to speak its own story of relationality in place, climate writers are urged to engage collaboratively with First Nations writers and communities.

The shortfalls in writing to celebrate cosy places, or to enjoin others to care for 'our place' in ways that imply ownership, are easily fallen for (Plumwood 2008, p. 143). A nostalgic or possessive view of place cannot alone answer global environmental concerns. Plumwood proposes, as do Donna Haraway (1988) and Ursula Heise (2008), that projects that focus on privileging place and assumed belonging to place have the potential to marginalise, deny or even displace dispossessed peoples and nonhuman entities.

In particular, cultural and ecological blindness, and ongoing colonial-type refusal to consider Indigenous forms of relationship to place have persistently limited place-based discourse as 'a genuine ecologically-aware discourse' (Plumwood 2008, p. 143).

For white settler writers, an important part of writing about Indigenous relationality with place entails seeking feedback from sensitivity readers. This enables writers to ascertain possible impacts of their work and to respond to a reader's feedback. First Nations readers must be approached sensitively, without an assumption that people have time or space to help out white-fella projects or writers from other cultures. Safe literary spaces should be honoured. Respectful exchange with Indigenous community members and communications with local Indigenous organisations, including seeking permission to represent Aboriginal and Torres Straight Islanders' stories, words or organisations, is essential (Wheeler 2013; Janke 2019). Meaningful relationships and authentic consultation with First Nations sensitivity readers are part of the efforts to 'get it right'. For example, portrayal of Indigenous characters includes understanding the importance of how people speak 'on Country', who can say what, and how familial and community relations impact the character's cultural connections. Taking responsibility and incorporating feedback from early sensitivity readers is an important part of writing that respects Indigenous connections to place. Learning of First Nations peoples' multidimensional connections to and ruptures from traditional custodianship includes striving towards cultural and ecological resonance in the tones of narrative groundwater.

Representation and Non-representation

A representational paradigm underpins much traditional nature writing and place writing. Representation of what is seen, heard, felt or sensed in fictional worlds and in creative non-fiction usually complies with traditional, 'common-sense' perceptions. From a representational paradigm, writers inadvertently assume their separation from an external world or object and that they can objectively represent a mimetic version of the world as they perceive it. Even as writers attempt to convey connection to place, writing representational texts inherently creates 'a cognitive separation between the world of the text and the world of ecological systems' (Harrison 2013, p. 1). Lawrence Buell's (2022, p. 4) critique of literature's representations of nature includes questioning assumptions of fidelity and giving insufficient attention to underlying human inadequacy in acknowledging social and cultural circumstances of environmental descriptions. Perceived binary separations between nature and culture, body and mind, human and more-than-human, rationality and affect have underpinned assumptions that humans can represent 'outside places' through various forms of communicative or artistic expressions. Nature writers have often assumed accuracy in their representation of places and things in the world. Climate writers face problems in representing the scale and temporal vastness of global warming in the Anthropocene. Likewise, the problems and politics of representing the quotidian impacts of groundwater's large-scale and long-term responses to global warming revolve around the problems of representation itself, as well as the impression that groundwater is at first invisible.

Representational language, while striving to create realistic images, has a propensity to occlude other possible meanings of the subject than the single observation or experience being presented (Hunter 1984, p. 415). Nature writing, in its literary expressions of place, often either hides or misrecognises the characteristics of language that make up descriptions of a place. Artistic works (painting, literature, music, sculpture) that endeavour to 'represent' place, particularly those that claim to create knowledge of or identity with a place, are likely to struggle with both correspondence and dissonance between the artist's/writer's ideas and material actuality. Representations of place aim to be evocative, but remain partial. In particular, the ideologies that underpin how the artist/writer 'sees' that place have a great impact on what is represented. Artists' and writers' task is to test 'the ideological assumptions we make about the world', particularly when the subjects we explore are otherwise represented as actual or factual by sexist and racist social norms or through the privileging of science and economic axioms (Millner 2016, p. 35). Choosing language that carefully navigates the social, cultural and political elements of place enriches the outcomes.

On both epistemological and scientific grounds, climate writers eschew representations of aquifers through fixed descriptions. Mimesis, or writing attempting to mirror reality, becomes a cul-de-sac for a climate writer's affective, ecological and political aspirations. With no single relation between language and the world, the task of writing from within the enmeshments of groundwater's far-reaching qualities and features is not one of fidelity or mimetic representation. Meaning making becomes less certain. Rather than trying to copy or imitate ineffable subterranean realms, I take heart in Paul Ricoeur's invitation to recast writing/reading as a re-description, 'a chain of interpretive processes' that will be constantly mediated through a myriad of values, experiences and beliefs of writers and readers (cited in Felski 2008, p. 84). Subterranean imaginaries are indeed chains of interpretations, within the heteroglossia[3] of influential voices surrounding the writer and reader.

Representational versions of 'reality' are contested under a broad umbrella of thought called *non-representation*. To partially address the problems of representation, climate writers in the Anthropocene increasingly challenge normative assumptions about what is known and how it is written. Degrees of coherence and/or disjointedness with the material worlds being investigated are acknowledged. Uncertainties of rendering unfixed, relational entities, such as groundwater, ask climate writers to express its plurality in multiple ways. Non-representational viewpoints are a means to overcome problems of mimetically representational expressions of places.

Non-representational theory has many permutations, which eclectically synthesise the sciences with philosophy, the arts and humanities in ways that cultivate understandings of the more-than-human, the more-than-textual, through events, practices and assemblages of everyday possibilities (Abram 1997; Thrift 2008; Vannini 2015b). With such breadth of scope, this approach to research and thinking appears to suit the fluidity and immensity of groundwater. A permeable exchange between the parameters of climate writing and non-representational perspectives provides an avenue into storing *with* an entity like groundwater, something that is not easily seen, but rather needs to be imagined.

Groundwater narratives cannot claim to make a singularly accurate expression of subterranean worlds because, beyond the reality of the text, the reality of a posthuman world is never fixed. Cultural theorists and philosophers have sought multiple ways to think and to write 'about' the world other than through a fixed lens. Realities are constituted through 'becoming-with' or 'wording' for Haraway (2016), through 'intra-action' for Barad (2017). For Morton (2013) hyperobjects, and for Latour (2017) the term *pluriverse* express the multiple realities that exist over time and across spaces. I maintain a claim that writing should however aim to convey realistically authentic worlds, as distinct from claims of mimetic realism. The excerpts from *Why we cry* attempt an initial exploration of the possibility that a realistic fictional world of becomings may be forged within a non-representational framework.

Climate writers may accept that the tones of groundwater in fictional worldmaking are inherently fallible. We may challenge to the possibility of accurately representing real world diversity. The movement of 'things' and their agency (two of Nigel Thrift's seven elements of non-representational theory; Thrift 2008) are concepts that assist thinking and writing that avoids describing fixed, external realities. In groundwater narratives, the three subjects, groundwater, the human protagonists and the author, are turbulently enmeshed with each other (Thrift 2008, pp. 9–13). Expressing groundwater's elusiveness, its unknowability, and conveying its relationality, becomes more important than persistently portraying its material characteristics (Vannini 2015b, p. 16). Writing climate fiction within the context of non-representational thinking entails considering one's porosity as a writer, and the myriad of influences seeping into the text.

In contemporary climate writing, writers non-re-present worlds of unknowable dimension, not just outside or alongside, but integrally entangled in the everyday world of human and nonhuman lives. Throughout the writing process climate writers develop a stronger sense of how groundwater is part of them, within them, not just adjacent to them. Groundwater narratives expand imaginary visions of their nonhuman subjects, aiming at the same time to elucidate how groundwater makes us human, to paraphrase Deborah Bird Rose (2000). Interrelationality prevails.

Ideas of soundscapes in writing about places provide another avenue for writers to expand non-representational aspects of place (Carter 1996, 2015). Soundscape ecology is a field of study that considers unique sonic collages of biological, geological and anthropogenic sites (Warren 2018). Related to documentary, film and sound media, Shilyh Warren advocates finding alternative voices to the 'anthropocentric voice of crisis' in non-traditional ecological soundscapes (Warren 2018, p. 101). Sounds of surface water can be garnered, as in the Erin Espelie film *Silent springs* that derives its title from Rachel Carson's book.[4] The tones of imaginary movements of groundwater are more difficult to capture or express. Recording sounds from deep underground requires specialist equipment. One example is the work by artist Lotte Geeven in Germany, who translated deep rumblings of the subterranean to a sonic range that can be heard by the human ear.[5] Seismography has recorded seismic noise to track water flows in aquifers, using sound waves as a method to map depths and flows of aquifers (Clements and Denolle 2018). Aggregating sounds as 'connective tissue' of nonhuman entities, as suggested in soundscape ecology, gels with efforts to quieten anthropocentric voices to enable voices of nonhumans to be heard (Warren 2018, p. 116). Warren's consideration of soundscape ecology emphasises 'the primacy and complexity of the voice as key to the aesthetic experience of inspiration – a vital element of activist practice' (2018, p. 116). Sounds of place, the metre of the land, can be reflected in the tones of words. To write groundwater as dynamic, never still and never finished, yet with movements that are 'audible' to the human ear, draws writers into another imaginary realm. Groundwater's imagined tones come to be expressed through soundscapes of words – deep and slow, oneiric musings of place. An inspirational example of a novel built on soundscapes is Kirsty Gunn's *The big music* (2012). It heralds and foregrounds sounds of Scottish bagpipes and hillsides through rhythmic repetitions of words, phrases and scenes. The novel circles like the *piobaireached* (pronounced *pe-brohh*), the music that is written to be played outside, music that cannot make itself small (Gunn 2012, p. xi). Cyclic iterations link sounds and people to place. The musical qualities of Gunn's novel suggest the dynamic qualities of the bare and windswept Scottish Highlands: the velocities of the wind, the bleakness of the horizons, the crazed meanderings of a senile old man.

In the following scene from *Why we cry*, Frankie converses with imagined aquifers, whose tones are lugubrious, doleful, slow. The words aim to evoke groundwater's lament.

Excerpt 5.2: *Why we cry*, Chapter 11

In the forest, beside the river, Frankie mumbled, 'I could construe a tale, write something down, as if it's real.' She walked slowly on a crackling dry pathway. 'You are real. You are more than language can construct. I make up words for you.' She sat beside a bulking eucalypt trunk, imagining stout-coloured porosities, grimy sedimentary air spaces below. 'You speak to me. I hear your deep voices and tremble. I have only sound waves uttered warily from my throat. Uttered warily, scarily uttered.'

The scratchy limits of words frustrated her. 'You have greater forces to convey your messages.' She lay on a bed of leaves, watching blueness through the canopy. 'I am tied to the urge to utter words, to etch them on a page. I am an addict to words.' She rolled over, furrowed lines in the earth with a stick. 'This is not how we converse. We are sensual together, delectable, smelly.'

She imagined deep bodies of water. Their replies descend, an avalanche of boulders vibrating timpani crescendos. You drain us. Down. Let us be. Down, down. Let us be.

She rested, spoke again into the cathedral-like bush chamber, slow and stuttering. 'Yearning tires me. So tired.' Her heart raced, stilled. Raced. Stilled. Fluttered like flags on a mast on a windy day. She felt tattered, scattered. She closed her eyes. Heard rustling in the canopy. 'Do you feel the wind?' She lay with her eyes closed.

'We are fading. Down, deeper down. Let us be.' Imagined subterranean bass-baritone tempers growled, howled, grumbled.

'Don't go.'

Groundwater's voices come to me in dripping half sentences, scribbled on bedside booklets, put there for the sake of not losing the semiconscious gems that arise from sleeplessness.

> As if we could talk. But if, if we could, it would be in deep growly grumbles, with solemn sibilance joining syllables. Lower tones than anything human-made, no explosive rumble for the profundity of our breathless groans. Chasms in the depth of a volcano's roar are growls above our baseline calls. We are on the move, unsettled, on the prowl. We flow.
>
> D. Wardle, unpublished notes, 2018

I collect sentences about imaginary sounds. What does the middle-of-the-night author do with these inspirations? What am I trying to express in these nightly invocations of groundwater's inarticulate sonnets – or are they somniferous in-articulations? These notes contributed to my subterranean imaginaries. Some made it into *Why we cry*.

Storying with narrative groundwater walks a tightrope between aspirations to present groundwater's materiality in realistic ways and the difficulties of expressing an invisible reality, the non-representability of groundwater. Contemporary stories that expose the political implications of human and nonhuman responses to climate change continue to steer climate writers and readers to new spaces between day-to-day narratorial tropes of realist, or otherwise-styled writing, and along convoluted, porous paths that merge threads of non-representational thinking. Creating language to express groundwater's fluid-filled places asks writers to compose narratives in groundwater's domain. The next chapter explores how writers and readers create affective and emotive dimensions in groundwater narratives.

Notes

1 Robert Scholes (1979) was an early literary critic to analyse the roles of metafiction and fabulation, as developments from traditional realist fiction. Scholes promoted the need to coordinate an author's efforts clarify ideas and emotional experiences in the novel form.
2 An overview of the story of the Kalapuya Stones can be found at Eugene Parks and Open Space (n.d.).
3 Heteroglossia is Bakhtin's (1981) term for the multiple influences that underpin literary imaginations.
4 See Espelie (2023) for more information about this 13-minute film produced in 2011.
5 For more information on Lotte Geeven's work see Stinson (2014). Also see Listening Biennial (2023) for thoughts on how deep listening can enhance storying for care-ful engagement with earthly voices.

References

Abram, D., 1997. *The spell of the sensuous: perception and language in a more-than-human world.* New York: Vintage Books.

Alaimo, S., 2008. Trans-corporeal feminisms and the ethical space of nature. *In:* S. Alaimo and S. Hekman, eds. *Material feminisms.* Bloomington, IN: Indiana University Press, 237–264.

Alaimo, S., 2010. *Bodily natures: science, environment and the material self.* Bloomington, IN: Indiana University Press.

Aronson, M., 2019. *Rising water: the story of the Thai cave rescue.* Cammeray, NSW: Simon & Schuster.

Bakhtin, M.M., 1981. *The dialogic imagination: four essays.* Austin, TX: University of Texas Press.

Barad, K., 2015. Transmaterialities: trans*/matter/realities and queer political imaginings. *GLQ: A Journal of Lesbian and Gay Studies*, 21, 387–422.

Barad, K., 2017. No small matter: mushroom clouds, ecologies of nothingness, and strange topologies of spacetimemattering. *In:* A.L. Tsing *et al.*, eds. *Arts of living on a damaged planet: ghosts and monsters of the Anthropocene.* Minneapolis, MN: University of Minnesota Press, 103–120.

Bawaka Country *et al.*, 2016. Co-becoming Bawaka: towards a relational understanding of place/space. *Human Geography*, 40, 455–475.

Bawaka Country *et al.*, 2018. Mobilising knowledges, identities and places as Bawaka. *In:* N. Gombay and M. Palomina-Schalscha, eds. *Indigenous places and colonial spaces: the politics of entwined relations.* Milton: Taylor and Francis Group, 51–71.

Bawaka Country *et al.*, 2022. Songspirals bring country into existence: singing more-than-human and relational creativity. *Qualitative Inquiry*, 28, 435–447.

Behrendt, L., 2003. *Achieving social justice: Indigenous rights and Australia's future.* Annandale, NSW: Federation Press.

Behrendt, L., 2021. *After story.* Chicago: University of Queensland Press.

Birch, T., 2015. *Ghost river.* St Lucia, Qld: University of Queensland Press.

Birch, T., 2016. Climate change, mining and traditional Indigenous knowledge in Australia. *Social Inclusion*, 4, 92–101.

Birch, T., 2020. Recovering a narrative of place. *Griffith Review – First Things First*, 60, 207–219.

Boulter, A., 2007. *Writing fiction: creative and critical approaches.* Basingstoke: Palgrave Macmillan.

Bowman, M., ed., 2015. *Every hill got a story: we grew up in country.* Richmond, Vic: Hardie Grant Books.

Buell, L., 1998. Toxic discourse. *Critical Inquiry*, 24, 639–665.

Buell, L., 2001. *Writing for an endangered world: literature, culture, and environment in the U.S. and beyond.* Cambridge, MA: Belknap Press of Harvard University Press.

Buell, L., 2022. *The environmental imagination: Thoreau, nature writing and the formation of American culture.* Cambridge, MA: Harvard University Press.

Carson, R., 1962. *Silent spring.* Boston, MA: Houghton Mifflin.

Carter, P., 1996. *The lie of the land.* London: Faber & Faber.

Carter, P., 2015. *Places made after their stories.* Crawley, WA: University of Western Australia.

Casey, E., 1997. *The fate of place.* Berkeley, CA: University of California Press.

Clarke, T., 2015. *Ecocriticism on the edge: the Anthropocene as a threshold concept.* London: Bloomsbury.

Clements, T. and Denolle, M.A., 2018. Tracking groundwater levels using the ambient seismic field. *Geophysical Research Letters*, 45(13), 6459–6465.

Coetzee, J.M., 2004. *Elizabeth Costello.* London: Vintage.

Contamination, 2017. Directed by Linton Besser. *In: Four corners*. TV, ABC TV, Oct 9.

Davis, M., 2003. Achieving social justice: Indigenous rights and Australia's future by Larissa Behrendt [book review]. *University of New South Wales Law Journal*, 26, 813–818.

De Santolo, J., 2018. Shielding Indigenous worlds from extraction and the transformative potential of decolonizing collaborative research. *In:* D. Fam, L. Nauser and P. Gibbs, eds. *Transdisciplinary theory, practice and education.* Cham, Switzerland: Springer International Publishing, 203–220.

Dillard, A., 1982. *Living by fiction.* New York: Harper & Row.

Emerson, R.W., 1836. *Nature* [online]. Available from: http://oregonstate.edu/instruct/phl302/texts/emerson/nature-emerson-a.html [accessed 3 February 2013].

Espelie, E., 2023. *Silent springs* [online]. Available from: www.erinespelie.com/cinema/silent-springs [accessed 20 January 2023].

Eugene Parks and Open Space, n.d. *Kalapuya Talking Stones* [online]. Available from: www.eugene-or.gov/DocumentCenter/View/3322/Talking-Stones-Brochure?bidId= [accessed 20 January 2023].

Felski, R., 2008. *The uses of literature.* Malden, MA: Blackwell Publishing.

Fuentes, C., 2004. *This I believe: an A–Z of a writer's life.* London: Bloomsbury.

Gay'wu Group of Women, 2019. *Songspirals: sharing women's wisdom of Country through songlines.* Sydney: Allen and Unwin.

Ghosh, A., 2021. *The nutmeg's curse: parables for a planet in crisis.* London: John Murray Publishers (Hachette).

Griffiths, T., 2018. Australian 'nature writing'. Conference paper. Environmental Writing: Creativity and Social Efficacy Symposium, University of Melbourne.

Gunn, K., 2012. *The big music.* London: Faber & Faber.

Haraway, D., 1988. Situated knowledges: the science question in feminism and the privilege of partial perspective. *Feminist Studies*, 14, 575–599.

Haraway, D., 2016. *Staying with the trouble: making kin in the Cthulucene.* Durham, NC: Duke University Press.

Harrison, M., 2013. The act of writing and the act of attention. *TEXT*, 20, 1–11.

Heise, U.K., 2008. *Sense of place and sense of planet: the environmental imagination of the global.* Oxford: Oxford University Press.

Hunter, I., 1984. After representation: recent discussions of the relation between language and literature. *Economy and Society*, 13, 397–430.

Iovino, S., 2014. Bodies of Naples: stories, matter, and the landscapes of porosity. *In:* S. Iovino and S, Oppermann, eds. *Material ecocriticism.* Bloomington, IN: Indiana University Press, 97–113.

Iovino, S. and Oppermann, S., 2014. Introduction: stories come to matter. *In:* S. Iovino and S. Oppermann, eds. *Material ecocriticism.* Bloomington, IN: Indiana University Press, 1–17.

Irigaray, L., 1985. *This sex which is not one*, trans. C. Porter with C. Burke. Ithaca, NY: Cornell University Press.

Irigaray, L., 2002. *To speak is never neutral.* London: Continuum.

Janke, T., 2019. *Protocols for using First Nations cultural and intellectual property in the arts.* Pyrmont, NSW: Australia Council for the Arts.

Johns-Putra, A., 2018. The rest is silence: postmodern and postcolonial possibilities in climate change fiction. *Studies in the Novel*, 50, 26–42.

Kingsolver, B., 2012. *Flight behaviour.* London: Faber & Faber.

Langton, M., 2006. Earth, wind, fire and water: the social and spiritual construction of water in Aboriginal societies. *In:* B. David, B. Barker and I.J. McNiven, eds. *The social archaeology of Australian Indigenous societies.* Canberra: Aboriginal Studies Press, 139–160.

Latour, B., 2017. *Facing Gaia: eight lectures on the new climatic regime.* Cambridge, UK: Polity.

Le Guin, U., 1984. The author of the acacia seeds and other extracts from the Journal of the Association of Therolinguistics. *In:* U.K. Le Guin, *The compass rose.* London: Panther Books, 11–19.

Le Guin, U., 2006. A very warm mountain. *In:* L. Savoy, E. Moores and J. Moores, eds. *Bedrock: writers on the wonders of geology.* San Antonio, TX: Trinity University Press, 146–153.

Listening Biennial, 2023. *The Listening Biennial 2023* [online]. Available from: https://listeningbiennial.net/ [accessed 20 January 2023].

Lucashenko, M., 2013. *Mullumbimby.* St Lucia, Qld: University of Queensland Press.

Macfarlane, R., 2019. *Underland: a deep time journey.* London: Penguin Random House.

Mahood, K., 2022 *Wandering with intent: essays.* Brunswick: Scribe.

Marshall, V., 2017. *Overturning aqua nullius: securing Aboriginal water rights.* Canberra: Aboriginal Studies Press.

McDonald, D., 1911. *The bush boy's book.* Melbourne: S.J. Endacott.

Millner, J., 2016. Preview: the 20th Biennale of Sydney: the future is already here – it's just not evenly distributed: the embassy of disappearance. *Broadsheet Journal*, 45(1), 33–36.

Mills, J., 2022. Dystopia creep. *Meanjin*, 80, 38–46.

Mitchell, E., 1945. *Speak to the Earth.* Sydney: Angus and Robinson.

Mitchell, E., 1946. *Soil and civilisation.* Sydney: Angus and Robertson.

Mitchell, E., 1958. *The silver brumby.* Sydney: HarperCollins.

Moggridge, B., 2010. *Aboriginal knowledge and cultural values of water.* Canberra: CSIRO.

Moreton-Robinson, A., 2015. *The white possessive: property, power, and Indigenous sovereignty.* Minneapolis, MN: University of Minnesota Press.

Morris, P., 2003. *Realism.* Abingdon: Routledge.

Morton, T., 2007. *Ecology without nature: rethinking environmental aesthetics.* Cambridge, MA: Harvard University Press.

Morton, T., 2013. *Hyperobjects: philosophy and ecology after the end of the world.* Minneapolis, MN: University of Minnesota Press.

Mulhall, S., 2009. *The wounded animal: J.M. Coetzee and the difficulty of reality in literature and philosophy.* Princeton, NJ: Princeton University Press.

Murakami, H., 2003. *The wind-up bird chronicle.* London: Vintage.

Neidjie, B., 1986. *Kakadu man.* Canberra: Mybrood.

Neidjie, B., 1989. *Story about feeling.* Broome, WA: Magabala Books.

Neimanis, A., 2017. *Bodies of water: posthuman feminist phenomenology.* London: Bloomsbury Academic.

Pascoe, B., 2014. *Dark emu.* Broome WA: Magabala Books.

Plumwood, V., 2008. Shadow places and the politics of dwelling. *Australian Humanities Review*, 44, 139–150.

Plumwood, V., 2009. Nature in the active voice. *Australian Humanities Review*, 46, 113–129.

Rose, D.B., 2000. *Dingo makes us human: life and land in an Australian Aboriginal culture.* New York: Cambridge University Press.

Rose, D.B., 2005. An Indigenous philosophical ecology: situating the human. *Australian Journal of Anthropology*, 16, 294–305.

Scholes, R. 1979. *Fabulation and metafiction.* Chicago, IL: University of Illinois Press.

Scott, K., 1993. *True country.* North Fremantle, WA: Fremantle Arts Centre Press.

Scott, K., 2011. *That dead man dance.* Sydney: Pan Macmillan.

Scott, K., 2017. *Taboo.* Sydney: Pan Macmillan.

Segal, M., 2017. The missing climate change narrative. *South Atlantic Quarterly*, 116, 121–128.

Sleigh, C., 2011. *Literature and science.* Basingstoke: Palgrave Macmillan.

Soper, K., 1995. *What is nature? Culture, politics and the non-human.* Oxford: Blackwell Publishers.

Spender, D., 1986. *Mothers of the novel.* London: Pandora Press.

Stinson, L., 2014. Listen to strange sounds recorded in a hole 5 miles deep. *Wired*, 9 January [online]. Available from: www.wired.com/2014/01/an-artist-records-the-mysterious-rumblings-of-middle-earth/ [accessed 20 January 2023].

Strang, V., 2013. Conceptual relations: water, ideologies and theoretical subversions. *In:* C. Chen, J. MacLeod and A. Neimanis, eds. *Thinking with water.* Montreal: McGill-Queens University Press, 185–211.

Thomashow, M., 2002. *Bringing the biosphere home: learning to perceive global environmental change.* Cambridge, MA: MIT Press.

Thoreau, H.D., 1993. *Walden and other writings.* New York: Barnes and Noble Books. (First published 1854.)

Thrift, N., 2008. *Non-representational theory: space, politics, affect.* New York: Routledge.

Todd, Z., 2015. Indigenizing the Anthropocene. *In:* H. Davis and E. Turpin, eds. *Art in the Anthropocene: encounters among aesthetics, politics, environments and epistemologies.* London: Open Humanities Press, 241–254.

Tredinnick, M., 2005. *The land's wild music: encounters with Barry Lopez, Peter Matthiessen, Terry Tempest Williams, and James Galvin.* San Antonio, TX: Trinity University Press.

Tree, I., 2020. Introduction. *Granta – Second Nature*, 153, 10–11.

Trexler, A., 2015. *Anthropocene fictions: the novel in a time of climate change.* Charlottesville, VA: University of Virginia Press.

Van Neerven, E., 2014. *Heat and light.* Chicago, IL: University of Queensland Press.

Vannini, P., 2015a. Non-representational ethnography: new ways of animating lifeworlds. *Cultural Geographies*, 22(2), 317–327.

Vannini, P., 2015b. Non-representational research methodologies: an introduction. *In:* P. Vannini, ed. *Non-representational methodologies: re-envisioning research.* New York: Routledge, 1–18.

Vermeulen, P., 2015. *Contemporary literature and the end of the novel: creature, affect, form.* Houndsmills: Palgrave Macmillan.

Warren, S., 2018. Hum, buzz, gurgle: ecological soundscapes in poetic eco-docs. *In:* M. Pramagiiore and R. Honess, eds. *Vocal projections: voices in documentary.* London: Bloomsbury Academic and Professional, 101–118.

Watt, I., 1957. *The rise of the novel: studies in Defoe, Richardson and Fielding.* London: Chatto & Windus.

Waugh, P., 1984. *Metafiction: the theory and practice of self-conscious fiction.* London: Methuen.

Wheeler, B., ed., 2013. *A companion to Australian Aboriginal literature.* Rochester, NY: Camden House.

Williams, R., 1975. *The country and the city.* St Albans: Paladin.

Wood, J., 2008. *How fiction works.* London: Vintage.

Wright, A., 2006. *Carpentaria.* Sydney: Giramondo.

Wright, A., 2013. *The swan book.* Artarmon, NSW: Giramondo.

Wright, A., 2016. What happens when you tell somebody else's story? A history of Aboriginal disempowerment. *Meanjin*, 75, 58–76.
Wright, A., 2022. The inward migration in apocalyptic times. *Meanjin*, 81, 57–65.
Wright, S., *et al.*, 2012. Telling stories in, through and with Country: engaging with Indigenous and more-than-human methodologies at Bawaka, NE Australia. *Journal of Cultural Geography*, 29, 39–60.

6

PERMEABILITY

AFFECT

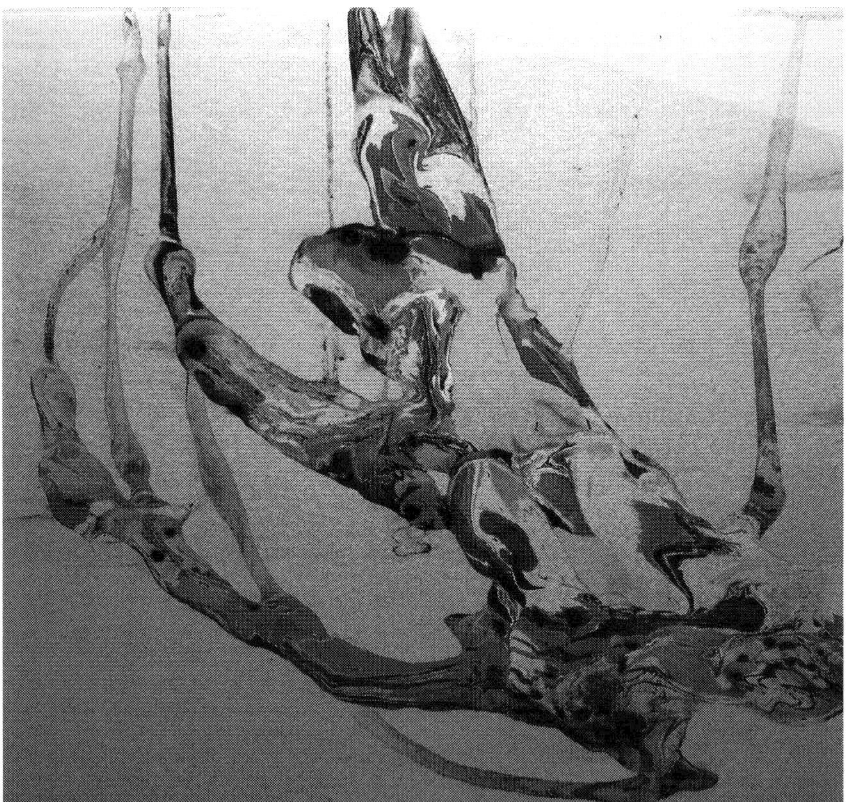

Figure 6.1 Bill Sampson, *Leap into the void 2023*, 2023, synthetic polymer on plasterboard as digital photograph.

DOI: 10.4324/9781003270416-6

The permeability of rock or soil is the degree that the pore spaces between the rock or sediment particles are interconnected. Geological permeability enables or restricts the movement of liquids or gases through various rock or stone materials. It is affected by how tightly the grains or fractures are arranged in porous materials. Permeability is also affected by the viscosity of the liquid and the pressure on any gases in the pore spaces. Permeability is slightly different to hydraulic conductivity, which is calculated by Darcy's law, a formula describing the rates of flux or discharge from a specified 'tube' or area, in a single direction through a permeable medium, over a specified distance and gradient. Permeability describes the rocky medium, rather than the flow rates through it.

> Attention without feeling ... is only a report.
>
> Oliver and Cook 2007

In this chapter I discuss how, in writing groundwater narratives, climate writers might express the affective capacity of an inanimate entity such as groundwater. The American nature poet Mary Oliver guides us towards the importance of expressing feelings and affects in groundwater narratives. I explore ideas about affect across four main themes: first the affective capacity of groundwater as matter; second the usefulness of thinking about osmosis as a form of shifting affects; third I explore haunting and amnesia as ways of generating affect in groundwater narratives; and fourth the role of writer/readers and reading in perceiving/conveying affect. Groundwater's subterranean flows through permeable rocks towards springs, into rivers or finally discharging in the ocean underpin its affective capacity. The potential for groundwater to be heard as an affective voice focuses our attention on its perambulating flows through permeable geologies. Groundwater flow provides an analogy for the seeping flows of human emotions either from internal places to the surface or to be hidden in deeper soulful terrain. The flows could be tears of humour, joy, sadness, fear, rage, shame or hope, among many responses. I write of watery, lachrymal repercussions, implying perhaps the grief of springs drying up, the fears of polluted aquifers or the joys of healthy groundwater-dependent ecosystems. I consider later in the chapter how the 'permeability' of readers to groundwater narratives might be affected, and the uncertainty of what affects groundwater narratives might have. The affective agency of aquifers to cause life and death, to instigate local and global impacts, fits with groundwater's positioning as 'vibrant matter' (Bennett 2020) as discussed in Chapter 2. The wider affective connotations of groundwater, its domain, its consumption, its relations, seep from its metaphoric potency. Groundwater bodies, being large, diverse, amorphous, invisible, sensible and subterranean, become an affective force, expressing their affects in multiple ways.

Entanglements of humans with groundwater, through our guts and our skins, our livelihoods, our souls and our imaginations, are equally various. It may be through sipping mineral water from plastic bottles; chewing on and ingesting hydrogenous plants that have been irrigated by waters drawn from deep aquifers; bathing and swimming in hot mineral pools that steam and smell of earthy sulphur; or talking with a local water boring contractor at the pub, watching his bitten fingernails and roughened hands absorbing condensation from his beer glass. The shared constituency of our bodies as bodies of water, as well as shared locations of humans and nonhumans, act as the conjoining affective forces in groundwater narratives. In *Why we cry*, a raft of characters show complexities of various enmeshments. Kasha Duff, the antagonist, sees groundwater as his property, as a resource. He sells groundwater for bottling; he schemes to provide water to a new housing development on his back block. Wade Keane is learning of the cultural significance of groundwater to his Indigenous community. Betty Winsome knows of groundwater's biblical history. Lennie Pankhurst feels for the water table through his divining sticks. Helen Mack stands firm for science, measuring groundwater's pollution plumes according to defined mathematical formulae. Each of these characters embed and embody affective relations to groundwater's ways of being. Each adds to the entanglement of readers' affective perceptions. Protagonist Frankie Pankhurst's perceptions and feelings preside in the close third-person narration. It is primarily through her thoughts, experiences and emotions that affective connections with groundwater's potency are performed. Anthropocentric viewpoints pull, wanting to influence the text. Groundwater's voices emerge muffled through human entanglements.

The urge to write with groundwater's voice pressed to the surface, springlike. I experimented with groundwater's 'dialogue' in first-person singular and first-person plural prose, seeking the affects of groundwater narratives.

> I weep, I seep, deeper than deep. I recede to darkness.
>
> We weep, we seep, deeper than deep. We recede to darkness.

I read and re-read the possibilities for affecting readers. Plurality resonated. Groundwater narratives emerged.

Excerpt 6.1: *Why we cry*, Chapter 8

As soon as she arrived home she collected her new ladder, carried it to the well. Puffing and sniffing, she pushed through the scrub, lowered the ladder into the gloom and climbed down to the calm of underground space. Billowing cumulonimbus clouds receded above. Saturated sandstones seeped. She closed her eyes, imagined hanging tendrils, roots of trees dangling in a subterranean cave. White-blind roots seeking moisture in cavernous air spaces, coiled in their search for water. Blood vessels withering in a moment of paralysing fear. Darkness prevailed. She sat on the bottom rung of the ladder, covered her eyes with her hands, considered the science that battered at her brain. Mathematical formulae can't guide water's flows. Numbers come from estimated models. It's all about how good the models are. Water travels broad and elusive trails.

Darcy's law. Henri Darcy found a sweet pleasure in mathematics. There's nothing wrong with that. Artois needed water, to think. We all need water to think. Hydration counts.

Frankie settled herself at the base of the well watching darkness, smelling the gloom. Thoughts and memories bumped into each other, like flotsam on a rushing river.

Rocks, spaces and water – together the holy triumvirate. Like people need a skeleton, need porous lungs for breath, and blood and the vessels to carry it. Images flashed of old bores, metal pump casings and long pipes. Open channels spewing artesian waters. It's all plastics now, tens, hundreds, thousands of metres of black tubing snaking across the land.

We are water. We are aquifers, moving, moved. We are water.

The mouth of the mineshaft turned gauzelike grey, she climbed the ladder to the surface. She decided to leave the ladder there to make it easier to visit the well. As she walked home a flock of cockatoos flew away, shrieking, dipping and turning, sulphur crests flared, looking to find a roost for the night.

Groundwater as Affective Matter

Worldwide, receding water table levels, and extended efforts to continue boring more deeply for water, show that 'groundwater is the most extracted natural resource on the planet' (Bessire 2022, p. 350). There are millions facing catastrophic impacts of aquifer depletion and water scarcity. Lucas

Bessire describes global patterns of aquifer depletion in semi-arid and arid regions:

> Today, the sites of extreme groundwater depletion constitute a distinct planetary topography. This shadow geography of empire and extraction is capable of structural adjustments of its own. Aquifer depletion may reshape terrain into its own image, reproduce hierarchies of access, and generalize certain experiences across time and space. In these places, the stakes of depletion are coming into sudden, sharp focus.
>
> Bessire 2022, p. 350

The notion of 'shadow places' was conceived by Val Plumwood (2008) to draw attention to those lost and unattended places that slip from cultural recognition to relational vacancy. Shadow places show the intersections of power and culture. Relating this concept to water, Jessica McLean and peers (2018) describe 'shadow water' as the water that is unattended, utilised at the whim of commercial powers with disregard for its many interwoven cultural and environmental enmeshments. Narrative groundwater becomes shadow water, for example, when the bureaucratic powers of groundwater management practices render a broader cultural and ecological view of aquifers as marginal or invisible. Groundwater narratives evoke the urgency of paying attention to groundwater's wider hydrological and cultural affects, beyond its capacity to be pumped for commercial gain. Narrative groundwater's affective potency bleeds not only into human lives, but also through complex nonhuman ecologies, expressed through multi-species entanglements. Groundwater-dependent ecosystems are examined in the next chapter.

Affect theory provides insights to understanding narrative groundwater's liveliness. Affects are non-cognitive and non-representational intensities outside of consciousness. Affects may be distinguished from emotions, or they may be seen as integrally linked with thoughts, emotion and feelings (Schaefer 2022). Elspeth Probyn (2010) and Sarah Ahmed (2014) argue, as did Deleuze, that making distinctions between affect and emotion, while initially useful, is not as important as seeing them as ideas that bring together 'collisions' and entanglements of both body and mind, of sensate feelings and thoughts. In feeling emotions, cognitive awareness of visceral responses is retained. Schaefer (2022, pp. 3–5) argues more cogently that 'thinking is feeling', dismantling the taken-for-granted, traditional divide between feeling and thinking. Emotions invoke a known interiority and boundedness to individual beings (Ahmed 2014; Weik von Mossner 2014, p. 1; Vermeulen 2015). Making distinctions between affect, emotions and feelings emphasises the autonomy of affects and their capacity to circulate between human beings and things, rather than being contained solely within human beings. The instant that humans perceive or are cognisant of an affect is the very moment that it becomes a felt emotion (Thrift 2008, p. 117). Nigel Thrift encapsulates how writers might express the resonances or affect of things, describing affect as 'the means by which bodies are guided along particular paths by things' (2008, p. 117). These may be human bodies and nonhuman things. Schaefer's argument focuses attention on how knowledge making is not only entangled with feelings, but is inherently embroiled in and draws on 'the same emotional aquifer' (Schaefer 2022, p. 3). In a similar way, how humans feel about groundwater will affect how they come to know of groundwater's impacts. Writing groundwater narratives entails an author's engagement with affect and emotion through their responses to groundwater's affects.

Attaching human meaning to narrative groundwater's affects is based on the enmeshments between cognition of emotions and intertwined affects. Affects that may be explored in groundwater narratives include threat, shame, fear, courage, hope or passions in politics.[1] I focus on the enmeshments between affect and cognition, the interplay of knowing and expressing felt emotions. This interplay is an important element of writing groundwater narratives that portray the potency of dissident voices of both human and nonhuman actants.

Building from the discussion in Chapter 2, efforts to change thinking from 'about' to '*with*' groundwater, towards an awareness of the inherent relationality of water, is aided by seeing the connections between affect and materiality. Jane Bennett (2010, pp. xii–xiii) considers affect as central to ethics and politics that work towards understanding and accepting the vibrancy of nonhuman matter. In Bennett's descriptions of 'impersonal affect', she ascribes affects as 'intrinsic to forms that cannot be imagined (even ideally) as persons' (2010, p. xii). Her focus is on the capacity of matter in and of itself to be affective and vibrant, rather than affect being an external force that might be added to matter. This approach to the inherent affective capacity of matter fits well with an approach to narrative groundwater that reveals the liveliness and affective capacity of aquifers. To apply the notion of 'lively' to a plummeting water table and depleted aquifers at first appears contrary. It is jarring to consider the catastrophic effects emanating from imperilled aquifers 'lively'. However, Bennett's 'lively' matter emphasises the affective capacity of matter, and opens the door to hope. Rewriting narratives to emphasise relational entanglements with water enables both humans and groundwater to be lively together.

Affect and emotion, in the context of these enmeshments, is explored in 'An inventory of shimmers' (Seigworth and Gregg 2010), the introduction to the influential text *The affect theory reader.*[2] Gregory Seigworth and Melissa Gregg outline the undulating territory of affect theory, enumerating at least eight possible approaches to how affect may be premised or understood. Building from a Deleuzian position, they interlace 'affect as an entire, vital, and modulating field of myriad becomings across human and nonhuman' (Seigworth and Gregg 2010, p. 6). Affects are located 'in the midst of things and relations (in immanence) and, then, in the complex assemblages that come to compose bodies and worlds simultaneously' (Seigworth and Gregg 2010, p. 6). Affect occurs in the enmeshments of complex assemblages between the human and the nonhuman world, as derived from a Deleuzian ethology of bodily capacities (originally from Spinoza) (Seigworth and Gregg 2010, pp. 5–6). Four of the eight overlapping and at times conjoined spheres of investigation of affect proposed by Seigworth and Gregg have relevance to thinking *with* narrative groundwater. First, affect may entail:

> Certain nonhumanist, ofttimes subterranean, and generally non-Cartesian traditions in philosophy, usually linking the movements of matter with a processual incorporeality (Spinozism): particularly as found in those contemporary approaches that try to move beyond various gendered and other cultural limitations in philosophy …
>
> Seigworth and Gregg 2010, pp. 6–7

The invitation to consider affect as part of a 'subterranean' way of thinking, giving focus to the generally unseen, nonhuman movements of matter in connection with a sense of incorporeality, fits well with a groundwater narrative's missive to express the affective capacities of aquifers. Groundwater's affects stem from a way of thinking, a philosophy, that indeed moves beyond the Cartesian separation of mind and matter, rationality and feeling.

Groundwater narratives, as we have seen in Chapter 2, build from a view of matter that give groundwater agential and affective capacities.

Second, affect may also be:

> Found in the regularly hidden-in-plain-sight politically engaged work – perhaps most often undertaken by feminists, queer theorists, disability activists, and subaltern peoples living under the thumb of a normativizing power – that attends to the hard and fast materialities, as well as the fleeting and flowing ephemera, of the daily and the workaday, of everyday and every-night life, and of 'experience' (understood in ways far more collective and 'external' rather than individual and interior), where persistent, repetitious practices of power can simultaneously provide a body (or better, collectivised bodies) with predicaments and potentials for realizing a world that subsists within and exceeds the horizons and boundaries of the norm.
>
> Seigworth and Gregg 2010, p. 7

Groundwater narratives aim to provide collective voices to the diverse forms and experiences of groundwater in many terrains. They give expression to the many interactions between aquifers and human and nonhuman collectivities. The politically engaged work of climate writing that draws upon groundwater's affective potency may, for example, attempt to illustrate the impacts of groundwater management strategies, where government and non-government power structures contribute directly to groundwater's circumstances. Advances in engineering have meant that deeper and deeper bores can be drilled to access water in deeper aquifers as the water table recedes. Numerous quantitative drivers increase extraction levels, including engineering of high-efficiency bore drills and larger submersible water pumps, inadequate regulatory controls and environmental protections, lack of research to continue to develop the hydrogeological knowledge base, population increases, and lifestyle choices in developed countries. Aquifers are receding. Stories about issuing excessive licences (beyond aquifers' recharge capacities), or lack of recognition of cultural connection to groundwater sites, or lack of regulation at all, are bound to challenge normative horizons. Such stories speak out for groundwater's affects. Human economic and political structures that take an extractive approach to groundwater are challenged in groundwater narratives that respond to and critique the debilitating effects of contemporary, repetitive actions of groundwater control systems.

Third, Seigworth and Gregg propose that another approach to affect:

> is located in practices of science and science studies themselves, particularly work that embraces pluralist approaches to materialism (quite often threaded through the revivification of Alfred North Whitehead's writings); hence, scientific practices that never act to eliminate the element of wonder or the sheer mangle of ontological relatedness but, in Isabelle Stengers's words, 'make present, vivid and mattering, the imbroglio, perplexity and messiness of a worldly world, a world where we, our ideas and power relations, are not alone, were never alone, will never be alone' (2007, 9). Here affect is the hinge where mutable matter and wonder (ofttimes densely intermingled with world-weary dread too) perpetually tumble into each other.
>
> Seigworth and Gregg 2010, p. 8

Groundwater narratives embrace the 'mangle of ontological relatedness' of groundwater's storying potency, through its inherent relationality. As has been previously discussed (Chapter 4) groundwater narratives' engagement with hydrogeology and other related groundwater sciences is imperative. Subterranean imaginaries are sites where science and the messiness of groundwater's material and affective impacts are tumbled together to generate creative works. Donovan Schaefer gives particular attention to ideas of 'messing together', saying: 'if reason is itself a passion, then it messes together with the other feelings that define our lives' (2022, p. 8). An example of generating an interchange between rationality and affect occurs in Barbara Kingsolver's *Flight behaviour* (2012). I return frequently to a key scene in this novel; the moment in one of Dellarobia's conversations with entomologist Dr Ovid Byron, where he explains that if the monarch butterflies die on Dellarobia's family block of forest over winter, they will not survive as a species. Dellarobia's reply is not an expression of 'words of science', 'but it was a truth she could feel. The forest of flame that had lifted her from despair, the migratory pulse that had rocked in the arms of a continent for all time: these fell like stones in her heart' (Kingsolver 2012, pp. 315–316). I feel the knocking in my chest when I re-read this scene. Kingsolver's fiction embodies an approach to science that never eliminates wonder nor entanglements of a messy world with impinging power structures. To express the potency of affect as a hinge between mutable materialism and awe remains the goal in climate writing. Groundwater narratives hinged with subterranean imaginaries aim to tumble groundwater's materiality with both wonder and mourning.

Fourth, I mention briefly that affect may also be seen as the psychobiology of innate, automatic, visceral, bodily responses to outside influences (Seigworth and Gregg 2010, p. 5). Human and nonhuman ecologies are deeply affected by loss or pollution of groundwater. Aquifer depletion or pollution causes and will continue to cause major disruptions. The story of the suicide in 2015 of Queensland farmer George Bender, in response to ten years of pressure from coal-seam gas companies to mine his property and threaten his water supply, could be the experience of many more. The insurmountable problems he faced from the lowering of the water table due to coal-seam gas mining in the district illustrate the seriousness of the threats to groundwater supply (Robertson 2015). Climate writers aim to embed the messiness of affect and emotion into groundwater narratives.

As I think about affect, I recall experiencing a range of feelings in writing emotional scenes in the fictional work *Why we cry*. The potential for exploring Deleuzian enmeshments of rationality and emotionality within the storying of groundwater emerged (Deleuze and Guattari 1988; Bogue 2010). Groundwater narratives have an important role to play in breaking traditional dualisms between the rational and the emotive by integrating cognition of the threats of climate change with human affective and emotional responses to changing environments. They seek to portray the significant effects of the many ways that groundwater affects human and nonhuman lives based on what we know about its characteristics and movements. Groundwater narratives emphasise conjunctions/enmeshments/entanglements of cognition and emotion to multiple expressions of groundwater's affective potency.

Osmosis and Permeability in Texts

The osmotic movement of water through and across semi-permeable membranes is affected by the chemistry of the fluid solutions on each side of the membrane and the pervious or impervious qualities of the membrane itself. Water wants to balance the concentrations on both sides of the membrane. It wants. Skin acts as a semi-permeable membrane, absorbing and excreting water and salts. Rock acts similarly. Groundwater moves through permeable geological materials as affects move through a text towards the production of meaning for the reader.

The following excerpt from *Why we cry* includes a brief exploration of interchange between human and frog skin, expressing porosity and permeability between human and nonhuman animals and their shared environments. Frankie holds a frog in her hand and feels its skin on hers. I aim to portray the affects of interspecies skin touching skin. As Karen Barad (2012, pp. 206–209) asks, how close are we when we touch, particularly when we let fall all pretence to separate the affective from the physics of touch. Movement between skins, the permeability of skins, stretches our imaginings of skins, stretches boundaries, and stretches humans out of a singular focus on ourselves towards a relational point of view. A parallel thread to the permeability of rock affecting groundwater flows is the permeability of skins, particularly frogs and human skin.

Excerpt 6.2: *Why we cry*, Chapter 11

'Wanna catch a frog?' Frankie asked Wade. Along the riverbank frogs called bass lines to the cricket sopranos; a symphony blasted.

'Sure, why not?' Wade crept to the riverside. Frankie followed. They perched, silent. Amphibious throats filled the summer air with vibrations. His calling, her silent approach. Frogs singing, surging, repeating their desires. Frankie listened to their love songs, the immediacy and yearning.

A frog's bulging eyes at the water's edge caught Wade's attention. With a snap, he cupped the frog in his hand. Their palms touched, their skins met. Its pale gulping throat pulsed.

'Got one.' Wade held his cupped hand towards her.

Holding the frog in her hands she felt the tender suction of balloon-fingered toes, the damp, cool belly skin, the taught musculature of legs ready to leap. She saw the pulsating throat, a pulse and breath, strong gulps of life. She was on a journey towards being more baffled, towards not knowing what she thought she once knew. Takes time and patience to develop a relationship with place, to get the knowledge, to understand rhythms and quirks of place, she mused, gazing at the frog. Connection to groundwater may take a lifetime, she thought mostly because its voices speak in innumerable tones.

'Choirlike', she muttered.

'What?' Wade watched her from the bank.

'Nothing, just thinking.'

Below them, aquifers spread, changed and moved in their own ways, like inkblots on textured paper. Slime slipped between Frankie and the frog, permeable, osmotic. Psychic skins, sticky skins. Their skins sensed the flimsiness and substance of emotions, the enclosed internal places where feelings circled, outside in, inside out. She leaned down and opened her hands. The frog leapt from her palm.

Permeable affects of the text aim to change readers' relations to frogs and to groundwater. As Sarah Ahmed argues, 'emotions create the very effect of the surfaces and boundaries that allow us to distinguish an inside and an outside' (2014, p. 10). The implication is to consider how humans are 'moved by feelings into a different relation to the norms that we wish to contest' (Ahmed 2014, p. 210). I discuss the potential links from feelings to agency and activism in the next chapter. Learning hydrological information is part of how writers feel about groundwater. Expressing emotional relationality provides an avenue to expand the metaphors for a connected life, a permeable text. Ascribing affective capacity to both real entities and imaginary subterranean places is foundational to the expression of groundwater narratives.

Amnesia and Haunting

How might a climate writer overcome the tendency to forget or ignore the value of healthy environments and strong groundwater reserves? Might amnesia or ignorance of groundwater's place and roles in ecosystems reduce its capacity to affect humans? The 'slow violence' caused by unfolding disasters with at times postponed effects, to use Rob Nixon's well-known phrase, asks us to confront, rather than ignore 'incremental and accretive' violences (Goodman and Shaikh 2018). Nixon says that the invisible phenomenon of climate change represents nothing less than an 'intergenerational theft of the conditions of life itself' (Goodman and Shaikh 2018). In a similar vein, the ecological term 'shifting baseline syndrome' describes the human tendency to forget how landscapes change, how species extinctions leave haunted amnesias (Gan *et al.* 2017, p. G6; Roberts 2020). Intergenerational changes in how humans perceive the world create new normals: 'Each generation sets its mental benchmark of normality by how the world looked when first encountered, often in youth, and see change relative to this. Younger generations accept as normal a world that seems tainted and degraded to older people' (Roberts 2020, p. 18). Unlimited access to groundwater may seem normal to previous generations, whereas contemporary, younger generations may see the limits to groundwater supply as normal from having experienced receding water tables and failing wells and bores. I am wary of generalising or falling for the trap of binary ways of seeing young and old, as of course attitudes to and recollections of groundwater may vary across a continuum at any age. Finding the traces of groundwater's actions in landscapes, recognising changes to groundwater-dependent ecosystems, becomes a way to reduce the losses of amnesia. Shifting baselines implies different losses; some will adapt and accept them, some will grieve, some will die from the lacuna-like absences caused by a receding aquifer. The slow violence of receding aquifers may not be dramatic at first, but will impact human and nonhuman species for many generations to come.

Could groundwater have memories? I propose that groundwater does not suffer amnesia; it carries recollections of its sources on its subterranean journeys in its accumulated chemical constituents, its viscosity, its capacities to navigate various permeabilities of its terrain. Groundwater narratives invoke the long histories and multiple slow violences of aquifer journeys.

To counter the tendencies for forgetting, following the groundbreaking work of Avery Gordon (2008), Elaine Gan and peers invite us to engage with the hauntings of ghost-like vestiges of landscapes that are 'imbued with earlier tracks and traces' (2017, p. G6). Avery Gordon says about haunting:

> What's distinctive about haunting is that it is an animated state in which a repressed or unresolved social violence is making itself known, sometimes very directly, sometimes more obliquely. I used the term haunting to describe those singular yet repetitive instances when home becomes unfamiliar, when your bearings on the world lose direction, when the over-and-done-with comes alive, when what's been in your blind spot comes into view.
>
> Gordon 2008, p. xvi

In a similar way, groundwater narratives invite readers to see and feel what has been in their 'blind spot', to feel unnerved by imaginings and realities of loss of groundwater. New planetary topographies will be understood through the lens of groundwater's hauntings, the gaps and presence that it leaves.

Ongoing over-extraction from aquifers will have devastating effects on human and nonhuman ecosystems long into the future. I imagine a failing or polluted aquifer as a ghost, a relic of earlier times, a haunting from past geologies, past deluges and soaking rainfalls. Subterranean imaginaries are not fantastical ghost stories; rather they bring to life the urgency of recollecting what would otherwise go unseen. Capitalist extractivism seeks to hide the ghosts of its profits. Cultural expressions of the ghost-like presences of ancient waters ensure that groundwater's memories are not forgotten.

The colonial project asks us to forget the devastations it has caused and the damages still occurring under the veil of its combined forces (Ghosh 2021). Colonial forces include ongoing racism, unjust distribution of wealth, inequities in gender and sexual preferences, dependence on consumerism. Capitalism relies on unsustainable economic growth that leads to increased fortunes of the wealthy at the expense of those who labour and are often forced to live in poverty.[3] Groundwater narratives invite storying *with* subterranean water sites to ensure recognition of groundwater's diverse social, economic and cultural potency. Amnesia holds its breath as time remakes a landscape under the forces of colonial values. Groundwater narratives may understandably have the tone of a lament. Climate writing, particularly expressions of narrative groundwater, limits the disappearances that come from ecological amnesias.

For those lucky enough not to suffer from environmental amnesia, those who recognise the intergenerational deficit of shifting baselines, or for those who pay attention to environmental histories and hear remnant, haunted voices, the losses over time are tragic. Receding aquifers threaten diverse groundwater-dependent ecosystems and unique environmental habitats. Many endemic species of plants, algae, mosses, insects, reptiles, birds, fish and mammals are at risk of being forgotten as they disappear. Groundwater narratives address the tragedies of loss.

Knowing predictions of further loss that is yet to be experienced is what Deborah Bird Rose (n.d.) refers to as the grief of the 'double death' – awareness of extinctions in time yet to come. As a writer of groundwater narratives, I suffer not only the possibility of groundwater's immediate 'demise'; I grieve its double death, knowing that the effects of loss of groundwater are not only occurring now, but are yet to happen to greater degrees, affecting millions of human communities and nonhuman ecosystems. Groundwater narratives expose the stories of 'shadow waters', the neglected subterranean entities that emerge to haunt us (McLean *et al.* 2018). Forgotten, depleted or culturally disregarded, groundwater voices are asking to be heard.

A risk in facing the challenges of the Anthropocene is that humans will continue to disregard nonhuman entities, will forget or ignore the evidence of anthropogenic impacts under the veil of idealised and idealising forms of nature writing. Groundwater narratives must be up to the task of breaking traditional assumptions that humans have no need to recognise their enmeshments with other forms of life and non-living entities. Jagodzinski expresses concern that new materialism's interest in vitalism needs to go further than 'an ethics of care, responsibility and attunement to the more-than-human, mapped against an "affective turn"' (2021, p. 586). The changes needed are more than how we map feelings. Responses to the Anthropocene, when material is considered at a quantum level (as by Barad and Deleuze), may not, according to Jagodzinski, sufficiently address the need for transformative social, political and economic changes in understandings of relationality with matter. Convincing readers that their responses only start with what they understand or feel is developed further in the next chapter.

Ascribing narrative groundwater with affective relational capacity implies mapping its stories at the level of both cellular and politico-economic responses. Emotions derived from storying *with* subterranean imaginaries may indeed involve, as Barad suggests:

> willingness to put oneself at risk, to place one's body on this wounded land, to be in touch with it, to have felt the sense of its textures, to come to terms with a shared sense of vulnerability and invisibility, to feel the ways this land, this void, which marks the colonisers' continuing practices of avoidance, always already inhabits the core, the nucleus of your being.
>
> Barad 2018, quoted in Jagodzinski 2021, p. 273

As bodies of water, human bodies are intimately linked with groundwater. Narrative groundwater speaks above human foibles: amnesias, appropriations of groundwater's voices, falling into dualist traps, avoiding the messiness of interrelationality, all fade beneath narrative groundwater's voices. The risks of distraction demand that writers look to the hauntings and traces of groundwater to ensure that humans do not forget subterranean voices. Narrative groundwater challenges a climate writer's sense of vulnerability and potential for failures. The illusive nature of aquifers and their affects are the aqueous hauntings that climate writers pursue. The urgent need for changed viewpoints in the Anthropocene spurs efforts to write into haunting silences, into spaces that need to be heard.

Affecting Readers: The Writer as Reader

Groundwater narratives enable the possibilities for affective and emotive responses from readers. Contemporary climate writers hand their work over to readers' 'interpretive responsibility' aspiring for the science and the narrative form to have impact (Sleigh 2011, p. 21). Coupled with listening to feedback from early sensitivity readers,[4] a writer writes and reads their own work with the aim to engage readers. Writers read as if their protagonists' experiences could happen to the reader. What happens when groundwater is the protagonist? How might groundwater itself affect readers? The importance of an affective politics to the work of writing climate fiction emerges.

The experience of reading enables readers to potentially understand new things and to feel emotion. This idea derives from Wolfgang Iser's viewpoint that, 'It is in the reader that the text comes to life' (1978, p. 19). Roland Barthes (2001) similarly gives life to the reader, as the author loses theirs. Analogies between groundwater's sub-surface movements and the below-the-surface emotions of writers/readers envisage new connections of feelings and cognition. Aspirations for climate writing to affect the readership prevail. Writers' intent seeps into the creative work. Writers aim to change readers' understandings and interests. We want readers to acknowledge enmeshments that they may otherwise feel disconnected from. Writers cling to the general concept that reading is thought to impart affect, through our use of imagery, through narrative, through language (Iser 1978; Trexler 2015, p. 71). Experimenting with varying perspectives from different narrator/s, different narratorial points of view, is a way to explore what voice will have the greatest affect, what voice will enable the reader to feel each character's emotions. Pieter Vermeulen encapsulates the leakage between emotions and affects: he describes 'the restless interplay' between writing of emotions and the 'affects that inevitably escape them' (2015, p. 9). Writers may hope for recognition, enchantment, knowledge and shock, the four elements framing the reader's approach to texts (Felski 2008). By recognition, Felski (2008, pp. 30–35) suggests, through text, readers can know something of themselves or the world that they may be either familiar with, or a stranger to. Groundwater narrative writers aim for readers to recognise something of the fragility of subterranean spaces, and the importance of recognising groundwater in relation to themselves and their emotions. Learning about groundwater's responses to the permeability of the rocks they occupy and pass through means expressing affective entanglements between groundwater, humans and nonhuman ecologies.

Discussing emotionality in texts, Sarah Ahmed explores the question 'What do emotions do?' (2014, p. 4). She argues that emotion is not 'in' the text, but operates as 'effects of the very naming of emotions' (Ahmed 2014, p. 13). Rather than assuming that readers' emotions are derived from texts, Ahmed (2014, p. 19) suggests that it is more useful to consider what the 'materials' are doing; how groundwater's actions, for instance, generate affect. As they find words and ways to engage readers in the emotions of groundwater narratives, writers ask readers to feel engaged with the ideas of groundwater's vulnerability and potency, its capacity to lead to life and death.

Readers' emotional responses to fictional characters and events are only partially distinguished from emotive responses to real-life experiences (Boruah 1988; Prose 2012). Readers can be in pragmatic and emotional relation to other people, events and places through their responses to texts (Pence 2004). A reader's empathetic understanding of fictional events, even to the state of tears, illustrates the 'reality' through imagination with which climate writing can be endowed (Boruah 1988, p. 215). Experiencing others, other lives, other entities, through reading climate writing, can be emotional and is part of the richness of an empathetic life (Boruah 1988, Ahmed 2014). Boruah (1988) argues that emotions generated by fiction are not 'fictional emotions', but valid expressions of empathy based on a combination of evaluative beliefs and imagination.

The ultimate affects in readers remain, however, speculative. Climate writers may consciously seek to engage readers in emotive responses through conveying fictional characters' emotions, but the affects emerging from writing remain unruly, and beyond authorial control. Climate writers hold onto the potential that readers may laugh and cry, may experience a sense of lament or sorrow at environmental losses, and/or be moved to consider groundwater differently as a result of reading groundwater narratives. Like the unimaginably leaky permeability of the terrain, I am unable to imagine (nor is it wise to try) the range of affective receptions by unknown readers at some time in the future. They are not knowable nor containable in the interplay between cognition and affect. Regardless of an author's writing/reading processes, the impacts on other readers is not in the text alone. It is in the range of complex beliefs, experiences and sentiments that readers bring to the reading that their aesthetic and affective experience is determined (Felski 2008, pp. 19–20). Writers will never know the prior perspectives or frameworks that readers may bring to groundwater narratives, based perhaps on cultural background, political values and life experiences. Has the reader ever watched a water diviner at work, seen spring water flow from a gleaming granite hillside, walked on the crusted algae of a dry watercourse? Neither the meanings nor the affects of the text are inherently fixed. They are negotiated through the relationships of various readers with the text. Ultimately, uncertainty lingers as to whether the creative work provides self-recognition, or recognition of relationality, whether the reading unsettles, shocks, or causes joy or rage to readers.

In the second phase of writing, after the first flush of inspiration is laid out, the author reads and edits their own work. Analysing for accrual of knowledge, and/or for how emotions might be felt through the text, it is essential for writers to read and assess their own writing. Being my own first reader, in the vein of Wolfgang Iser's (1978) ideas of reading as an 'aesthetic response', means I am the first to consider the affects of the text, initially through my embodied responses. It is my recognition of the bodily affects of my words on me the writer who has become the first reader/writer of the disappeared author's work that spurs the narration onwards. Donna Lee Brien describes this as a 'second type of reading'; reading for a writer/reader becomes a 'speculative and imaginative' activity (Brien 2006, p. 56). The writer/reader is reading for not only pragmatic coherence, but also for the affect and the evocation of emotion in the work. The reading may indicate to the writer, from their own responses, the possibilities that other readers may become affected by the work.

Reading as a writer occurs like an 'absorption' process, not dissimilar to osmosis, where liquids flow through a semi-permeable membrane from places of low to high concentration, aiming for balance. In this analogy, ideas move across a permeable membrane consisting of all the encumbrances that a writer brings to the writing/reading process. I bring my brain and intellectual perspectives, my heart and its propensity for emotion, my experiences coloured by veils of memories. My own watery and osmotic cells are all part of the writing/reading process. Political aspirations for changes in knowledge and attitudes to groundwater and to groundwater management regimes also come into play. As a climate fiction writer my 'observational reading of life' (Brien 2006, p. 55), the insights and analysis I apply, also affects how I collect, express and analyse materials and ideas in my writing. As my own first reader, I feel my own responses to the text I have created; I notice its affects on me.

Throughout the drafting processes, while aiming for readers to feel affective engagement, climate writers become aware of their own emotional responses to various scenes. Writing as a creative practice is inherently proximal to the concept and practice of reading: 'indeed writing reduces to be so indistinguishable from the moment of reflection or self-reflection in which I look back at what is now the text … the act of reading can seem to be what produces the writing' (Harrison 2013, p. 3). Writers' own responses to the story becomes an important place of decision making for how storying *with* groundwater evolves. I cry when Frankie farewells her father. I smile when Frankie possibly found groundwater with her father's dousing sticks. I punch the air when Wade successfully influences the outcome in the VCAT hearing. Frankie Pankhurst's journal entries enable her to reveal a similar self-awareness as a writer/reader, which readers likewise may recognise. They become a gesture of her authorial, self-reflective voice. Re-reading her tattered teenage journal, she thinks of it as 'self-absorbed claptrap'. She is a tough critic of her own writing, as I am of mine.

As a climate writer/reader reads their own work it is essential to listen for disharmony in authorial voice. Maurice Blanchot (1999, p. 460) raises the question of intrusions of the author, the voice speaking in the background, prompting characters and events with either authority or complacency. He warns of 'an authoritarian or complacent "I" still anchored in life and barging in without restraint' (1999, p. 460). Blanchot, who focuses on fiction, says that 'the novelist is a person who refuses to say "I", but delegates that power to other people', those 'little egos' who speak for us (1999, p. 461). Climate writers become self-conscious readers, aspiring at the same time to shuck invasions of egotistical hankering. Roland Barthes also argues that, for a fictional work to be effective, and affective, 'the voice loses its origin', the author is gone, the identity as author 'slips away' in the moment of writing (2001, p. 1466). However, when authors read their own work, simultaneous reading of the work illustrates authorial presence in the text. The duplicitous, slippery terrain occupied by a climate fiction writer/reader prevails. The duplicity is partially deceptive – as a first reader a climate writer might attempt to deny or ignore their own authority and intentions, listening for the polyphonic voices from the text. At the same time, the affects of reading our own writing reverberate bodily. The double-ness implied in the act of writing/reading refers to the reverberations that permeate between the affective and cognitive processes of writing-reading-re-writing-re-reading. Diminution of the writer's ego ensues, as language struggles with its limitations, as expressions of the narrative become those of other beings and entities, characters and voices outside the author's control. According to Blanchot (1999), the quieter the writer's authorial voice, the more clearly 'characters' – human and nonhuman – speak.

Which of my middle-of-the-night conversations with aquifers made the muster? As I was writing *Why we cry* I was drawn towards expressing my protagonists' 'conversations' with groundwater, in the manner that Angela Rawlings (2016) describes 'conversations' as metaphors for mutuality in human-to-landscape encounters. A shared language comes from sensual materiality; meaning that humans and more-than-humans are deemed to be capable of reading and writing the signs of the 'semiosphere' (Rawlings 2016, p. 268). Respectful communication, or conversations, with wide human and nonhuman kinships model new forms of dialogue and responsible entanglements (Rose 2005, 2013). Deborah Bird Rose describes dialogue as 'situated availability', a 'necessary concept for the slow work of ethics and witness' (2013, p. 7). In novel writing, Bakhtin (1981) calls these conversations 'dialogic imagination'. Groundwater gives few opportunities for situated or physical observations when it dwells underground. It is an imaginary conversation I have; it is a sensing of affects I read for. I chose at times for Frankie's dreams to be the carriage of her emotion, of her dialogues with groundwater. Her regular journal entries added another way to show her conversations with a subterranean imaginary.

Excerpt 6.3: *Why we cry*, Chapter 5

She dreamt long and complex dreams. Riders on horses coming, galloping, endless galloping towards stunned and waiting crowds. Why don't the people move, wave their hands or something. Perhaps they're trying not to be seen. Catastrophe will hit them regardless. The horsemen. There are no horsemen. They have all gone out to pasture, removed their weighty saddles and jingling bridles from broad backs and sweaty heads of noble equines. They've put down the cudgels and now lie resting on calm green meadows, blowing dandelion seeds, wondering what all the fuss was about. There is no catastrophe coming, they seem to say, as they whistle and hum in the sunshine. They are merely resting. Waiting to catch us all off guard. They're just trying to fool us. Even the horses are lying on the grass, noses resting on sweet earth, their tails curled over folded hooves. Their eyes are calm, so knowing. There is no danger in the wind, no threat of drought, no chance of suffocation from sands and dust. They all look so relaxed.

A white coated, bespectacled giant strides across the hillside, muttering.

They don't see the monster wave coming. Run, run away from these dreaming deceivers. The voice of a newsman drones on.

You're a panic merchant, someone shouts.

A voice bellows. Save the aquifer. Save the aquifer. Save the aquifer and you save life. A crowd of voices swells; flags flew on the hillside horizon.

Frankie woke from her dream, trembling, wondering whom she'd been talking to. Where had her words come from? She reached for her journal, let new words flow:

> *Some say nothing bad is really going to happen. So what if we have a few dry years? So what if it's only a drought that draws down the aquifers, and large, deep-rooted trees wither and crack, thump to the ground, no longer able to reach their water source? That's what happens when you're taking a closed-eyed, shadowy approach. You lose sight of the dangers. Who's doing the pumping, the polluting then? Who will forget the drought, the fallen tree, the lost habitats?*

She bit at the end of the pen, watched shadows of branches of the tree tapping at her window. Cast by moonlight they wavered across the floor of her room. She continued to write:

> *You can't point at an aquifer. Could you paint one? I don't know what I feel. It's all bunched up with loss and hope, tragedy and hope. Always hope. How can I forget hope?*

Water traders take the cake. Well might they splash out and enjoy each purchase, as if there's no tomorrow. They will see you blotted out in their wealth. Purchasing power swallows the poor.

Taking, taking, taking the road to death. Death isn't resting; death is waiting to pounce.

Just because we can't see artesian water doesn't mean that it's not doing something, that it's not a source of life, the foundation of large and complex ecosystems. I reckon its best to leave it alone, to connect and intermingle in all its wondrous ways. Ways we could hardly imagine. Or could we?

She chewed at her pen again. Her thoughts wandered into a blank space. And from nowhere words sprang again:

Aquifers need someone to talk for them. Who can best do that? Not those horsemen, whoever they were?

Guilt wrestles with paralysis, paralysis leads to shame. Do something, you ninny.

I can't talk for underground water if I'm running fearfully from imaginary horsemen. See them there, resting, giving you a chance to get a word in.

Speak up, you say.

I have no voice.

You must speak. You are water itself; you must speak.

This fight is not fought by individuals. We will shout together.

Through Frankie's self-reflexive commentary in her journal I aimed to show the conjunctions of thoughts, dreams and awareness of possible forgetfulness. I wrote then read the draft. I felt as uncertain as ever that I had created a world in which readers will hear groundwater's voice. Meaning making through affect rests ultimately in the imaginaries of readers, as they 'write' in their own minds what they are reading (Barthes 2001). The initial instance of this process commences with my first reading.

I wrote/read/re-wrote/re-read, in one instance understanding that the text I had produced was momentary, insufficient and ultimately in the hands of unknown readers. In the next moment I wrote with heart-stopping trepidation of this very unknown, and under a cloud of anxiety that the manuscript *Why we cry* may not convey the ideas it had the potential to generate. The potential for shame quickly followed. As Elspeth Probyn explores, the bodily experience of shame derives from terror that my writing is insufficient to the scope of my subject (2010, pp. 73–74). She suggests that the blush from reading early writing should compel writers to keep writing with a desire to strengthen the work (2010, p. 89). Despite my fears of inadequate expressions of groundwater's potency and vulnerability, I wrote with a moral compass, honestly grappling with my own emotions, hoping to engage others in the affects of underground water. In the early reading and editing of my fictional work, I applied criteria of aesthetic, ethical and political impact. I brought a feminist critique of androcentric assumptions. I brought my cognition, as well as my emotional disposition, to the task of reading. My first reading may still miss all these marks. Efforts to make affective stories from the ephemeral nature of groundwater go beyond authorial control. The 'specific explosion of mind, body, place and history' is Probyn's (2010, p. 81) description of the subjective entity of affects such as shame, fear and anger. The 'explosion' becomes the yardstick with which a climate writer reads their drafts and explores the relationships between themselves/the writer, the textual expressions, and their possible affects. Climate writers enter terrain of multiple messy entanglements, where the life of the work is reliant on wide research and deep exploration of interactions between inside and outside worlds, interior emotions and complex networks of affects.

Through the processes of writing, reading and re-writing, groundwater's affects materialise in the reader's body. An interplay emerges between what readers might come to know about groundwater and how that knowledge might affect them. Groundwater narratives illuminate the inherent feelings of human enmeshment with groundwater, aiming to induce any range of emotions in readers. They might lead to joy, rage, determination or tears. There is no hierarchy within the range of emotions that may be generated.

The narrative conjunction of knowledge and affect, alongside a diminution of authorial voice, aims to lessen the possibilities of moralistic pedagogy. I do not dodge the political responsibility of climate writing that attempts to move readers to places of new knowledge, to new affective and political responses to the potency of groundwater in the global water crises. By drawing out emotions in narrative groundwater, writers aim to open readers to new thinking and actions.

This chapter has argued that the precarious materiality of aquifers can be expressed to readers through the enmeshments of knowledge making and the affective capacities of humans and nonhuman alike. Juxtaposing or interweaving rationality and affect, knowledge and feelings through the rigours of writing/reading/rewriting strengthens what groundwater narratives might 'do'. Writing groundwater narratives that matter, stories that move readers to new understandings, stories that engage them with groundwater's multiple affects is timelier than ever. Within the pluralities of climate writing forms (poetry, fiction, creative non-fiction, video games, film and theatre scripts, etc.) groundwater narratives give expression to the affective becomings of a shy and receding subject. How these efforts translate to potential literary agency and activism is discussed in the next chapter.

Notes

1 In the burgeoning affective turn numerous scholars have addressed these topics: threat (Massumi 2010), shame (Probyn 2010), fear (Ahmed 2014) or passions in politics (Amin and Thrift 2013). There is not the scope here to analyse each of these possible affects.

2 Greg Seigworth and Carolyn Pedwell's *The Affect Reader 2* (2023) reorients affect theory to contemporary worlding and affective expressions.

3 For an overview of capitalism's contribution to the Anthropocene, and its impact on 'nature', see John Bellamy Foster's *Capitalism in the Anthropocene: ecological ruin or ecological revolution* (2022). Foster (2022, pp. 141–152) gives an overview of the problems of the term 'nature', and the injustices in remaking colonial territories according to colonists' imaginings, through the Marxist lens he applies throughout the book.

4 Sensitivity readers read for specific knowledges and cultural beliefs. In groundwater narratives they may be First Nations readers, hydrogeologists or water lawyers.

References

Ahmed, S., 2014. *The cultural politics of emotion.* Edinburgh: Edinburgh University Press.

Amin, A. and Thrift, N., 2013. *The arts of the political: new openings for the left.* Durham, NC: Duke University Press.

Bakhtin, M.M., 1981. *The dialogic imagination: four essays.* Austin, TX: University of Texas Press.

Barad, K., 2012. On touching – the inhuman that therefore I am. *Differences: Feminist Theory Out of Science*, 23, 206–223.

Barthes, R., 2001. The death of the author. *In:* V. Leitch, ed. *The Norton anthology of theory and criticism.* New York: W.W. Norton & Co., 1466–1470.

Bennett, J., 2010. *Vibrant matter: a political ecology of things.* Durham, NC: Duke University Press.

Bennett, J., 2020. *A vitalist stopover on the way to a new materialism.* New York: Duke University Press.

Bessire, L., 2022. Aquifer aporias: towards a comparative anthropology of groundwater depletion. *Current Anthropology,* 63, 350–359.

Blanchot, M., 1999. *The Station Hill Blanchot reader: fiction and literary essays.* Barrytown, NY: Station Hill.

Bogue, R., 2010. *Deleuzian fabulation and the scars of history.* Edinburgh: Edinburgh University Press.

Boruah, B.H., 1988. *Fiction and emotion: the study of aesthetics and the philosophy of the mind.* Oxford: Clarendon Press.

Brien, D.L., 2006. Creative practice as research: a creative writing case study. *Media International Australia,* 118(1), 53–59.

Deleuze, G. and Guattari, F., 1988. *A thousand plateaus: capitalism and schizophrenia.* London: Athlone Press.

Felski, R., 2008. *The uses of literature.* Malden, MA: Blackwell Publishing.

Foster, J.B., 2022. *Capitalism in the Anthropocene: ecological ruin or ecological revolution.* New York: Monthly Review Press.

Gan, E., *et al.*, 2017. Introduction: haunted landscapes of the Anthropocene. *In:* A.L. Tsing *et al.*, eds. *Arts of living on a damaged planet: ghosts and monsters of the Anthropocene.* Minneapolis, MN: University of Minnesota Press, G1–G14.

Ghosh, A., 2021. *The nutmeg's curse: parables for a planet in crisis.* London: John Murray Publishers (Hachette).

Goodman, A. and Shaikh, N., 2018. Rob Nixon: gov't inaction on climate change is 'slow violence' that hits world's poor the hardest. *Democracy Now!,* 2 August.

Gordon, A.F., 2008. *Ghostly matters: haunting and the sociological imagination.* Minneapolis, MN: University of Minnesota Press.

Harrison, M., 2013. The act of writing and the act of attention. *TEXT,* 20, 1–11.

Iser, W., 1978. *The act of reading: a theory of aesthetic response.* Baltimore, MD: Johns Hopkins University Press.

Jagodzinski, J., 2021. Quantum creativity: afracting new materialism in the Anthropocene. *Qualitative Inquiry,* 28, 586–595.

Kingsolver, B., 2012. *Flight behaviour.* London: Faber & Faber.

Massumi, B., 2010. The future birth of affective fact: the political ontology of threat. *In:* M. Gregg and G. Seigworth, eds. *The affect theory reader.* Durham, NC: Duke University Press, 52–70.

McLean, J., *et al.*, 2018. Shadow waters: making Australian water cultures visible. *Transactions of the Institute of British Geographers*, 43, 615–629.

Oliver, M., & Cook, M-M., 2007. *Our world*. Boston, MA: Beacon Press.

Pence, J., 2004. Narrative emotion: feeling, form and function. *Journal of Narrative Theory*, 34, 273–276.

Plumwood, V., 2008. Shadow places and the politics of dwelling. *Australian Humanities Review*, 44, 139–150.

Probyn, E., 2010. Writing shame. *In:* M. Gregg and G. Seigworth, eds. *The affect theory reader.* Durham, NC: Duke University Press, 71–90.

Prose, F., 2012. *Reading like a writer: a guide for people who love books and for those who want to write them.* London: Aurum Press.

Rawlings, A., 2016. Ecolinguistic activism: how and why to rite. *In:* S. Herbrechter and E. Friis, eds. *Narrating life: experiments with human and animal bodies in literature.* Leiden: Brill, 267–302.

Roberts, C., 2020. Shifting baselines. *Granta: Magazine of New Writing – Second Nature*, 153, 13–25.

Robertson, J., 2015. George Bender's wife attacks mining lobby for claim his death was 'hijacked'. *The Guardian*, 28 October [online]. Available from: www.theguardian.com/australia-news/2015/oct/28/george-benders-widow-attacks-mining-lobby-for-claim-his-death-was-hijacked [accessed 11 November 2015].

Rose, D.B., 2005. An Indigenous philosophical ecology: situating the human. *Australian Journal of Anthropology*, 16, 294–305.

Rose, D.B., 2013. Slowly – writing into the Anthropocene. *TEXT*, Special Issue 20.

Rose, D.B., n.d. Double death. *The Multispecies Salon* [online]. Available from: www.multispecies-salon.org/double-death/ [accessed 16 November 2022].

Schaefer, D.O., 2022. *Wild experiment: feeling science and secularism after Darwin.* Durham, NC: Duke University Press.

Seigworth, G.J. and Gregg, M., 2010. An inventory of shimmers. In: M. Gregg and G.J. Seigworth, eds. *The affect theory reader.* Durham, NC: Duke University Press, 1–25.

Sleigh, C., 2011. *Literature and science.* Basingstoke: Palgrave Macmillan.

Stengers, I., 2007. Diderot's egg: Divorcing materialism from eliminativism. *Radical Philosophy*, 144, 7–15.

Thrift, N., 2008. *Non-representational theory: space, politics, affect.* New York: Routledge.

Trexler, A., 2015. *Anthropocene fictions: the novel in a time of climate change.* Charlottesville, VA: University of Virginia Press.

Vermeulen, P., 2015. *Contemporary literature and the end of the novel: creature, affect, form.* Basingstoke: Palgrave Macmillan.

Weik von Mossner, A., 2014. *Moving environments: affect, emotion, ecology, and film.* Waterloo, Ontario: Laurier University Press.

7

GROUNDWATER-DEPENDENT ECOSYSTEMS

LITERARY ACTIVISM

Figure 7.1 Bill Sampson, 2020 *Bushfire detritus*, 2020, digital photograph.

DOI: 10.4324/9781003270416-7

Groundwater-dependent ecosystems (GDEs) are the complex, natural, biological life systems that require access to groundwater for some or all of their life cycles. GDEs require groundwater to maintain the ecosystem processes sustaining diverse communities of plants and animals. GDEs rely on both surface waters (such as ephemeral pools and swamps directly fed by groundwater) and subterranean waters. The hyperaeic zone, the plume-like area beneath creeks and rivers, and deep subterranean aquifers support unique life forms. A wide range of vegetation from algae to water plants, rushes, reeds, shrubs and trees rely on shallow groundwater in specific locations. Vertebrates such as fish, frogs and reptiles often survive in springs and waterway pools that are fed by groundwater. Invertebrates, from many types of insects in surface pools to microbiological life forms, including minute crustaceans and stygofauna that live in subterranean aquifers, also rely on groundwater flows. GDEs support both localised and landscape-scale ecosystems. Limited research on the interrelationships between surface water ecologies and groundwater ecosystems means little is known about the diverse biological niches groundwater supports. As the water table drops following excessive groundwater extraction, and as aquifers are polluted, rare and endemic species of plants, fish, frogs and invertebrates living in isolated and highly specific groundwater sites are increasingly at risk of extinction.

> We are creative, sensitive, imaginative humans that can use art to inspire a shift in the emotional world not only of ourselves, but of others.
>
> Gergis 2022, p. 211

This concluding chapter proposes that storying *with* nonhuman entities enables their material agency to be expressed as part of social responsibility and 'ethical relationality' (Todd 2015, p. 249). I discuss how climate writers may create human and nonhuman relationality in their writing, producing 'an enactment of ecological imagination' (Donald, quoted in Todd 2015, p. 249). Later in the chapter I discuss the role of climate writing in linking relationality with affect, agency and activism. Joëlle Gergis (2022) arrives at a point of illumination as a scientist, when she realises the power of literature and the arts to contribute to the cultural evolution, or rather revolution, needed to shift the negative trajectory of climate change. As well as changing emotional worlds through groundwater narratives, I explore the importance of acknowledging climate writing as inherently embedded within echelons of cultural and political power, and circumscribed with social and ethical purpose to inform and affect readers (Bradley 2017; Andersen 2020; Cole 2021). Erik Swyngedouw (2004) draws attention to the social power of water, focusing on inequities in water distribution and activism in Santiago de Guayaquil, Ecuador. He writes:

> All these grass-roots mobilizations and actions raise the issue as to who has the right to water. Whose water is it? And for whose city? ... Yet these grass-roots mobilisations and forms of protest have to be understood against the backdrop of the urban political institutions and the choreography of political power.
>
> Swyngedouw 2004, p. 156

His analysis of the parallels in flows of power and water is one of many examples where activism of various types operates within the 'choreography of political power' (2004, p. 158). More and more scholars, novelists, poets and creative non-fiction writers are challenging inadequate responses to environmental catastrophes caused by climate change.[1] Activism against political inertia and failures to mitigate further catastrophic global warming takes many forms. Writing as a form of climate activism highlights sustainability and environmental justice, and in the main gives focus to positive futures. Social and environmental justice underpin stories of activism and hope.

Groundwater's silence in many ecological debates and in most climate fiction often leaves me feeling that I have a 'protagonist' who is reluctant to be heard, reluctant to find its place in its own story. The precarity of many groundwater-dependent ecosystems (GDEs) can be seen as akin to climate writers' unique explorations of groundwater's place in the Anthropocene. As we have seen, exploring different expressions for groundwater's active enmeshments and affects, a climate writer considers epistemology, storying and science to shape and figure groundwater's narratives. This final chapter draws connections between literature's agency and environmental activism in climate change texts.

Biodiversity within ecosystems reliant on groundwater systems is poorly understood in comparison to surface waters (Gibert and Culver 2009). Groundwater-dependent ecosystems exist in subsurface aquifers, around isolated waterholes along watercourses, and in springs, wetland soaks and swamps that rely on the slow seep of groundwater through dry times to maintain unique life forms (Brodie *et al.* 2012; Geoscience Australia 2018). GDEs may occur many kilometres from a visible water source. They are nourished by subterranean water flows. The watery cells of multifarious animal and vegetative lives and the microbial biodiversity of GDEs are an important foundation to larger, landscape-wide ecosystems and food chains (Pain 2005; Eamus and Froend 2006; Griebler and Lueders 2009). Limited research by biologists and bio-hydrogeologists has meant there is limited information about the uniqueness of groundwater-dependent ecosystems, and in particular the fragility of unique flora and fauna that rely only on groundwater (Humphreys 2008). With added pressures on aquifers from the effects of climate change and population growth, GDEs provide important territory for groundwater narratives.

Life forms in isolated GDEs may be endemic to each specific locality. For example, the Dalhousie hardyhead (*Craterocephalus dalhousiensis*) lives only in the tepid waters at the sacred site Irrewanyere, part of the system of mound springs across South Australia and the Northern Territory. The Dalhousie hardyhead is found nowhere else on the planet. Irrewanyere is the Indigenous name for Dalhousie Springs, a tourist and camping area in the northern reaches of South Australia. Other extremely rare freshwater fish, such as the barred galaxia, survive only in remnant waterholes of mountain waterways, reliant on connections to aquifers to maintain water levels. Fish and amphibians are the most critically endangered species in freshwater ecosystems as many rely on groundwater to maintain habitats in stream and riverbed pools through dry periods (Cresswell and Murphy 2016). Little-known primitive groundwater stygofauna (crustaceans, mites and worms) live in the interstitial water-filled spaces of various rock formations. A group of stygofauna called Bathynellidae are so small (often less than 1 mm) that some live in between grains of sand in freshwater aquifers. Bathynellidae occur in aquifers worldwide, but in Australia only one has been fully described, in Western Australian aquifers, *Bathynella primaustraliaensis*. Requirements for environmental surveys related to mining activities have meant that an increasing number of subterranean taxa are being identified through complex sampling of existing bores (Perina *et al.* 2018). Many unique species of subterranean fauna are at risk from mining activities and excessive groundwater extraction (Pain 2005; Perina *et al.* 2018). Lack of knowledge of the biodiversity and distributions of subterranean fauna limits the scope of ecological impact assessments (Perina *et al.* 2018). Trees and smaller plants that depend on their roots reaching through the hyporheic zone to the water table in large wetlands risk death from either drowning or withering droughts. In many geological formations, excessive irrigation raises the water table and drowns trees. Excessive extraction lowers the water table beyond usual seasonal fluctuations, meaning that some trees die from thirst. Subterranean imaginaries touch upon the fragile nature of groundwater-dependent ecosystems. GDEs that spawn unique and diverse life forms become fertile analogical and real sites for climate writers to draw upon.

Boring for and pumping groundwater to make it accessible for domestic or industrial use, such as mining and agricultural consumption, is escalating as engineering and drill bit technologies allow for deeper and more efficient drilling. As populations increase, as agricultural demands for irrigated foods increase and as droughts become more prevalent in arid and semi-arid regions, over-extraction of groundwater has become a global problem. The lasting environmental impacts unique to each extraction circumstance are not always known. As was discussed in Chapter 1, human interventions are changing the balance that groundwater contributes to the global water cycle.

Literature's Agency: Capacities to Induce Change

Climate narratives describe social or environmental change in fictional worlds and/or in the real world. Chapter 2 discussed ways of knowing groundwater that impute its liveliness as agential matter. A human or nonhuman actant (originally Bruno Latour's term) is a source of actions, a force that catalyses events (Bennett 2010, p. 9). Agency may be connected to humans and nonhumans. Jane Bennett calls it 'thing power' (2010, pp. 4–14). Supported by Bennett's project to consider the distributive agency of human and nonhuman actants, groundwater narratives aim to 'direct sensory, linguistic and imaginative attention toward a material vitality' that contributes to social and environmental justice (Bennett 2010, p. 19).

At the same time, it is imperative to balance evocations of material agency with accountability of specific humans for environmental damages of the Anthropocene. Giving voice to nonhuman entities must not diminish a focus on human responsibilities. Such accountability should never legitimate retribution or vengeance, nor condone violence. Attributing agency to human and nonhuman actors aims to contribute to social and environmental justice as artists and writers seek to express ethical relationality in their work. Redressing the long-term silencing of Indigenous voices in art and literature remains a significant project.

Groundwater narratives give credence to the aspiration that literatures with embedded politics remain an effective agent of potential change (Felski 2008; Mulhall 2009). Like bodies of water, a body of literature can do many things. It disperses information, ignites new thoughts, produces emotions and affects. The potential for climate writing to contribute to social impetus for climate action rests in literature's claim to motivate agency. Groundwater's actions and enmeshments with human and nonhuman lives embed subterranean imaginaries as acts of resistance to long-held binary, colonial, capitalist thinking. Tuning into the entanglements of human and more-than-human agency in climate writing includes constructing otherwise unheard voices of the more-than-human world. In groundwater narratives this creates momentum towards the agential forces of literature.

Instigating change through literary works however, often faces institutional and cultural resistances. Rachel Carson and her book *Silent Spring* (1962) faced disparagement and attempts to discredit the information conveyed. I quote in full here the prologue and response from Charmaine Papertalk-Green and John Kinsella's collaborative poetry collection, *False claims of colonial thieves* (2018). They express the problems of creating change when white privilege and easily sold ideologies in schools deafen our capacities to listen to the land, to the needs of Country, to the needs of subterranean ecologies, and to whole water systems which nourish life. Too often we fail to hear the stygofauna speak.

Prologue

The stakeholders want their environmental scientists to deliver
flora and fauna on a platter, and they will do so for a price.
Stygofauna speak up through the land; some listen, more don't.
And so the mining companies reach into our schools,
funding programs that make students in their own image,
filling the holes they make in country with propaganda
sold as learning, gatekeeping into the church of university.
JK

Prologue Response

Privilege blindness
if environmental scientists say so
water comes from a plastic bottle
what lies on or within country
cannot be seen for the
privileged are privilege blind
CPG (Papertalk-Green and Kinsella 2018, p. xi)

Groundwater narratives must surmount such pernicious challenges.

Climate change's broad and multiple effects, its interdependent local and global predominance, as well as the race for currency of scientific and technological solutions in a rapidly changing world, has meant a paucity of system-wide and political responses to climate catastrophes, particularly a lack of regard for social justice. The damage of climate change is predominantly being experienced by populations and environmental entities that have least contributed to global warming. Michael Segal (2017, p. 124) argues that climate change narratives remain contested on many fronts and broadly do not meet the mark of instigating radical change. This joins Amitav's Ghosh's (2016) oft-cited complaint that contemporary climate literatures are not addressing 'the great derangement' of the Anthropocene. For Ghosh the great derangement is the continued insufficiency and inequity of real and literary responses to climate change. Yet, literary imaginations have, in the years since Ghosh bemoaned their dearth, significantly addressed escalating climate catastrophes. The large-scale and widely dispersed problems of global warming are increasingly being written into climate narratives that aim to quicken the hearts and minds of diverse readers, including potentially world leaders. However, climate writing still needs to expand its representation of stories that embolden efforts to create social justice. Climate narratives must address social, racial and environmental injustices as they create aspirations for change and emphasise cooperation over exploitation and competition. This is an important strategy to widen the scope of climate writing. Climate writers aim to contribute to a wide social and environmental justice movement through and beyond our writing. Storying *with* the agential materiality of groundwater coheres with social and environmental justice principles, by giving voice to 'things' that are infrequently heard.

By engaging groundwater's affective and agential presence, writers aim to be truly present, to be entwined, enmeshed in the 'myriad unfinished configurations of places, times, matters, meanings' (Haraway 2016, p. 1). Climate writers from all cultures write to create bridges that traverse inclusive perspectives, rather than attempting to mirror singular realities that have historically been riddled with colonial, patriarchal limitations. Groundwater narratives expand the agential capacities of groundwater's enmeshments, aspiring to enact ecological imaginations founded on ethical responsibility.

An example of Australian fiction that links the affects and agency of a river with the activism of the human characters is Anson Cameron's novel *The last pulse* (2014). Here the actions of a flooding river underpin what the protagonist aims for. This is a rollicking tale of a man who takes individually motivated, drastic action to address the loss of water flow from the Darling River, through the Murray River to his orchards in South Australia. Bombing a dam on a huge cotton property in Queensland (a thinly disguised Cubby Station) to release the last possible pulse of water flow down the Darling River through to South Australia offers readers a clear portrayal of a contemporary water activist in Australian fiction. Cameron's portrayal of desperate feelings in his protagonist, an unlikely water activist, Merv Rossiter, stems from Merv's grief over the suicide of his wife, and his reckless pursuit of water for his farm. After bombing the dam wall, Merv rides the wave of released water down the inland rivers in a stolen boat, accompanied by three unlikely people: his young daughter, an Aboriginal boy and a politician. They become a captive audience to Merv's views and actions. Climate writing, such as that embodied in Cameron's novel, that includes descriptions of the tragedy of failing river systems makes an excellent contribution to the hydrologic turn. In addition to a focus on surface or river waters, groundwater narratives include evocations of the interactions between aquifers and surface waters. Cameron tells a story of a man's individual decisions and actions, including 'surfing' the released wave of water down the river. Groundwater narratives might also explore the cooperative processes of water activists, showing their capacity to speak out, to lead their communities in their resistance to damaging groundwater drilling. Activism is rarely an individual pursuit. Cameron's novel threads Indigenous viewpoints and the complex political dilemmas of water management in the Murray–Darling Basin.[2] The pulse of the river is strong throughout this novel, and inspires writers looking for ways to show the pulses of groundwater bodies, as they respond to gravity, moon cycles, to floods and droughts, to pollution and human drilling and extraction. Expressing an activist's voice directly through a human protagonist is one strategy to give a focus to affect and agency.

Climate writers aspire for their stories to lay seeds, to spark, to instigate not only new thoughts and feelings but also a sense of agency and connection to inspire effective climate change responses. We hope to inspire courage and for readers to engage in climate activism. Potentially restraining this aspiration, Matthew Schneider-Mayerson's (2018) early research on the impact of reading climate fictions found that cli-fi readers do not in the main have a stronger engagement with environmental concerns after reading climate fiction. Schneider-Mayerson initially found that: 'The affective responses of many readers suggest that most works of climate fiction are leading readers to associate climate change with intensely negative emotions, which could prove counter-productive to efforts at environmental engagement or persuasion' (2018, p. 473). To partially explain this, Hennessy and colleagues comment on the prevalence of dystopian climate stories, arguing that 'there is a worry that the hopeless narratives most prevalent in cli-fi instigate feelings of despair which preclude activism' (Hennessy *et al.* 2022, p. 3). This implies that there may be a connection between reading texts that focus on catastrophes and apocalyptic scenarios and feeling despondent, and consequently less likely to engage in climate activism (Cole, 2021, p. 2). This could suggest, alternatively, that writing narratives of hope and successful change, not simply of optimism, could have greater potential to invoke propitious responses in readers. Am I clutching at straws? Schneider-Mayerson *et al.*'s (2020) more recent research suggests that climate writing that transports the reader to new 'places' and enables identification with new perspectives can have small effects on readers' ideas and beliefs. Groundwater narratives that bridge realistic climate change scenarios and relational thinking become important avenues for tapping the potential to bring about changes in perceptions. Climate change activism, including literary activism, takes more than optimism; it requires courage, resourcefulness, accurate knowledge and connection to community as a start.

Community-based politics and activism have been effective in drawing public attention to environmental damage caused by a wide range of industries including petrochemicals, agriculture, forestry, paper production, metals and plastic manufacturing, pharmaceuticals, computing hardware; the list is endless. In his discussion of 'toxic discourse', Lawrence Buell traces the development of environmental activism from preservationist agendas (efforts to preserve 'wild' or 'natural' places mostly led by wealthy and middle-class white males) to environmental justice campaigns (often against toxic waste dumping and often led by women and 'non-elites') (Buell 1998, p. 642). Social and environmental movements in the twenty-first century are increasingly in interplay in what Buell calls 'eco-justice activism' (1998, p. 643). The environment movement has battled relentlessly against governments worldwide, which continue to side-line and delay effective environmental law and policies. Sustainable and clean energy, supported by government policy and environmental agencies, is starting to raise the bar in select industries: for example, sustainable building practices, removal of fossil fuels from the energy system and transport systems, sustainable agriculture practices and plastic recycling. Such actions often miss the aim of ecojustice activism, failing to conceptualise environmentalism as an instrument of social justice (Buell 1998, p. 643).

The paradoxical gap between knowing of an impending crisis and having an ability to act to prevent it perhaps relates to the chasm between knowledge and beliefs (Baucom and Omelsky 2017). Widespread human inaction on climate change may be based on the disjuncture between what humans might know about climate change and how we feel about possible responses. A range of cultural, political and economic factors work against effective human responses to escalating climate-driven catastrophes. On an international scale, for example, after years of avoidance the United Nations Climate Change Conference, COP27, in Egypt in November 2022 negotiated a breakthrough agreement to provide 'loss and damage' funding for vulnerable countries hit hard by climate disasters (United Nations Climate Change 2022). Some argue that the time taken to settle this agreement detracted from urgent decisions on improving mitigation targets and ensuring a limit of 1.5° warming remains a global target (Morton 2022). To cross the chasm between knowledge and beliefs to actions feels like an interminable leap. Climate writing aims to boost this leap, guiding readers to know and feel ethical relationality in the material effects of global warming, and to be inspired to contribute to actions at whatever scale or locale is possible.

Aiming to bring about environmental, social or political change as a creative writer is in many ways an immeasurable aspiration. Roland Barthes's (2001) proposal that texts are 'written' by readers opens possibilities for the agency of readers to be acknowledged. Authors continue to hope that their climate writing increases readers' interests or capacities to make a positive difference to the worlds they inhabit. Julie Sedivy (2017) argues that reading literature can influence the mind state and empathy of readers. She says, 'When an author expresses deep confidence in a reader and creates a space in which the reader can, from the depths of her own social imagination, lower her consciousness into the body and experiences of another, the effect can be transformational' (Sedivy 2017). Climate writers may not know for sure who reads or listens to their stories, poems or essays. What myriad of effects or affects the creative work evokes may include transformations, or it may not. The movement from knowledge, through affect and belief, to political action is under-researched in environmental literature (Schneider-Mayerson *et al.* 2020). Ecocritical analysis in relation to climate literature expands scholarly readers' understanding of literary affects, but as yet research is sparse about the pathways from affect to activism for a wide readership. As Schneider-Mayerson *et al.*'s (2020) more recent research indicates, beliefs and attitudes may be influenced by reading climate fiction, but the likelihood of translating these to specific actions such as voting patterns or participation in environmental campaigns is not yet determined.

Ecocritical film studies have examined various film genres to uncover environmental films' impacts on viewers' emotions and understandings of environmental problems. The research shows that film's 'relevance to human experience and attitudes to nonhuman nature have occupied, at best, a marginal place' (Weik von Mossner 2014, p. 1). Eco-cinema, similar to eco-fiction, has consciousness-raising and activist intentions in relation to planetary health as primary goals (Weik von Mossner 2014, p. 2). Film and literary scholars recognise that emotion and rationality are intertwined as persuasive forces. Questions remain as to whether strong and meaningful affects and emotions presented in eco-films or climate writing, and effectively felt by viewers or readers, lead to environmental activism. Climate writers may seek to write with dissent, to shine a light on one small effect of climate change, but we will never really know the effects of our work on readers' thoughts, feelings or actions. The research is still being done. Readers, it can only be hoped, become motivated to join others in fighting for the environment's cause. Groundwater narratives aim to provide readers with strong political traction and links to current potential for activism.

The solitary and often depressing task of writing about climate catastrophes and social injustices is a difficult one. Hennessy and colleagues (2022) suggest that climate writers need to work collaboratively, to support each other through explorations of what are often bleak topics. Engagement with like-minded writing communities is an important way for climate writers to develop resilience in addressing the escalating problems of climate change. Climate writers may also join aspirational activist movements that link social concerns with finding solutions to environmental problems, as a way to sustain motivation to contribute literatures that make a difference. Writers and activists of all ages contribute to climate actions through organisations such as: in Australia, SEED, an organisation for young Indigenous people contributing to sustainable actions on environmental issues, particularly the problems of fracking for coal-seam gas in the Northern Territory; the Australian Youth Climate Coalition works for climate justice on several campaigns including national campaigns to keep fossil fuels in the ground and to support sustainable energy sources; the global School Strike for Climate movement also effectively engages young voices; Grandmothers for Climate Action; and in the USA the Elders Action Network.[3] People of all ages continue draw attention to the need for action on climate change. Literature's task is to support them in their work.

Kaleidoscopic views of various human actors colour and shade different ways that groundwater may be perceived: the water trader in a high rise tower making money from buying and selling water rights, the politician who seeks to secure votes with unworkable water plans, the Indigenous traditional custodian of water catchments seeking to maintain cultural water rights, the dairy farmer or potato or corn grower reliant on bore water to irrigate crops, the manager of a water authority mandated to issue groundwater drilling licences, the car wash operator or spa complex manager who is reliant on bore water for their business, the hydrogeologist concerned with paucity of data, the environmentalist who is outraged by loss of groundwater-dependent ecosystems. The list goes on. Each perspective reveals different human investments in access to groundwater. It is also vital to hear the voices of nonhuman players in groundwater-dependent ecosystems: the platypus colony facing extinction as waterholes dry up, frog populations that lose breeding habitat, the eucalypt standing gaunt, grey and skeletal in parched wetlands or flooded irrigation paddocks. Where human actions are causing the demise of GDEs, multifarious groundwater narratives are more important than ever. Making fictional expressions of groundwater, making storied meaning from groundwater's performances and its multiple affects, leads climate writers to address complex socio-political and economic issues through a range of stakeholders.

Finding authentic voice is a subplot of *Why we cry*. Frankie Pankhurst's journey to water activism starts with developing the courage to speak at a local council meeting.

Excerpt 7.1: *Why we cry*, Chapter 11

Frankie and Lois moved to the front row at the community hall. A faded photo of Queen Elizabeth hung on the cream-painted wall; the varnished honour roll glistened golden lists of names of young men from the district who had died in various wars. The Save Our Groundwater banner drooped lazily from its pins, defying the energy that went into painting the bright blue letters. Frankie listened to people coming in behind her. She glanced around: seats were filling fast. Clem and Wade sat either side of Lizard Green and Betty. Helen and Kim leant together on the edge of their chairs. Farmers held their hats on their laps. Workers from the factory nodded at Frankie as they shuffled along aisles. Don and Clive, Pam the bookkeeper, Bev the crop grower, and several familiar faces from the pub took their seats. At the back of the hall, Kasha Duff stood with his hat pulled low. Bernard bobbed beside him. Frankie scanned the room, imagining for a second that she might see Lennie's face, come to hear her. Typical, she thought, he still comes and goes.

The crowd stopped scraping chairs on the wooden floor, let out a few last coughs, hushed. Local councillor Susan Sheridan stood at the mic, tapping its end as if she was giving it heart resuscitation. The room fell quiet. Melody Blayne straightened the pile of folders in front of her on the desk, prepared to take the minutes of the meeting.

'Hullo and thanks for coming out this afternoon. Some of you know your bores are running dry, bores that have always filled with seasonal rains. Most of us know that the drought is the worst we've had. Some want this town to grow. Others are worried about the consequences.' Councillor Sheridan paused.

'We've had droughts before, and they'll come again', a voice creaked from the back of the room.

'It's about how we keep our town alive', a voice from the crowd interjected.

'How do we keep the water in our river?' someone else called out.

Sheridan continued. 'This meeting is about Mr Kevin Duff's housing development application. Council has received a petition, and objections to the planning permit application by Mr Kevin Duff ...'

'Call him Kasha. Cashed-up Kasha.' A shrill voice provoked chortles and sighs.

'Tell us why our tanks are dry', another voice called.

The room smelled of diesel fuel and perfumes, something smouldering before a gust starts a fire.

Councillor Sheridan ignored the remarks and went on to describe Kasha's application for the permit to drill for water. The water authority had approved the quantities of water that Kasha had applied for, and now council was considering the community objections, and whether to proceed with the permit for the works. The council engineer sat sweating at the table, head down, taking notes. 'Council is interested in how we keep our town alive. Yes, and we want to keep the water in our river. But we can't make it rain.' She glared at the crowd. 'You need to tell your local councillors what you want in terms of new developments, use of common resources. We all own the river. While it flows the town has hope.'

'Groundwater ain't common, it belongs to ...' Kasha's interjection was silenced by Councillor Sheridan's raised hand. Frankie turned to see Kasha shaking his head, scowling.

'I believe that Frankie Pankhurst has been nominated by the objectors, Save Our Groundwater, to speak, to summarise the material detriment that objectors may experience if the permit to drill is approved. You might remember ...' Councillor Sheridan was cut off by a call from the crowd; a thin man in a drooping Akubra cried out, 'Let her speak. What has Lennie Pankhurst's girl got to say?'

Frankie had endured the introduction with a flushed, lowered face. She stood, took the mic from Councillor Sheridan, holding her speech notes in a trembling hand. She began reading, then left off the page, letting words flow.

'Last December my dad died, then his horse died of thirst. Made me want to know why. I'm a beginner at this groundwater stuff, but there's something important going on. Aquifers around here are fairly short, kinda narrow, mostly fed by a deep lead, from an ancient riverbed, about sixty metres below the current surface. They're fed from the high lands, south of here. There's a lot of geology in the sixty million years that it took for the aquifers around here to form. Some of them come from ...'

'Get to the point. We know our land, we know how deep we drill', a woman called.

'Draw-down is like when the longest straw in a cup gets all the drink. Leaves the shorter straws stranded. Around here, it's not just new deep bores being drilled, it's the principle of people taking too much water for individual profit, without a concern for the effects.'

A farmer stood, spoke out. ‘I use groundwater for my potatoes. You going to stop me doing that?’

‘Are you on the side of those bureaucrats, bloody politicians? You wanna steal my water’, another farmer called across the room.

Sheridan stood, held her arm aloft. ‘There’ll be time for questions at the end.’

Frankie continued. ‘But it’s not just water, it’s lots of things. We could take a wider view.’ She took a long breath and continued in a lower tone. ‘Water and energy are connected. Our loss of groundwater is probably connected to a lot of things, including global warming. So we might do more than just think about saving our groundwater. Though this is pretty important around here. Everything you do makes a difference.’ She paused and looked at the bunch of faces looking at her. It felt like looking into a bowl of goldfish where by chance they all turned to look up at her together.

‘Water and energy are connected’, she repeated. She paused, looked at her notes, folded them and put them in her pocket.

‘I want to encourage you to take a chance, the chance of acting not just for yourselves, but for this community. Start by talking to your neighbours. There’s a lot we can do, a lot we must do. Don’t think you are in this alone.’ Frankie knew her face was red, her throat dry and her voice wobbling with emotion. Speaking out was not acting, in the sense of an actor acting out a role, creating a shell around who they really were. She remembered Lois saying that audiences recognise empty words. Speaking out was not just about the words that came from her mouth, but about her being genuinely herself, carrying all her experiences, all her psyche in her words. Suddenly she knew what Lennie had meant when he said, ‘Go to the source.’ She could only hope that the audience heard her, detected her authentic presence.

She continued slowly, nervously, turned to the heart of her message.

‘We live in Cawling not just because it’s cheap. I live here not just because I was born here. My connections are to the natural things that make this environment. They’re nothing compared to fifty thousand years of occupation of the local mob. Our job is not only to speak up for Country, on behalf of various creatures, or rivers and forest or groundwater. It’s already speaking for itself. We have to listen to what they are saying to us, in all their ways.’ She paused to try to moisten her mouth. A fan in the hall swirled warm air slowly around the crowd. Frankie continued.

'It's under stress. We might try to render in words what we see and experience around us: the air, rocks, fire and water. This is not just talking about the weather. We have to shout out the importance of what we know. Science tells us what is failing – the effects of global warming are everywhere.' She took a deep breath, saw Lois nodding, she kept on.

'What's all this got to do with groundwater?' Another wave of impatience flared.

'She's trying to steal your groundwater.' Sounded like Kasha Duff's voice, gravelly, determined.

'What's this got to do with groundwater? Well may you ask. Excuse the pun.' Frankie paused, someone giggled, others dug elbows into their neighbours' sides. 'It's not me who's stealing it.' She looked directly at Kasha. Bernard bobbed faster. 'Unfortunately mostly we don't see groundwater, it's invisible, unless you're a hydrogeologist.' Frankie nodded at Helen, who beamed encouragement at her. 'It doesn't take much to know that we are running out of water.' She cleared her throat.

'What you going to do? What are we meant to do?' Another heckling voice.

'Our essential connections, particularly to water, that stuff that makes us, must surely make us want to speak out, for the sake of preservation. For all of us. And by us, I mean everything – starting here with water. Think of the river. Pretty quickly I think of the plants and animals, calling to us in so many voices.' Frankie stopped as she knew her propensity to move onto the remarkable nature of frogs would lose some of her audience. Not all could relate to descriptions of the juicy, permeable zone occupied by amphibians. 'There's a planning permit application before council for a housing development down on Quencher's Road, near the river. An eighty-metre-deep bore. It will use close on two hundred megalitres of water every year. Water that's already spoken for, by the river and by those of us who already rely on that water for domestic bores. I urge you to object to council. I urge council to refuse the permit for twenty-five new houses. We don't need them. We need replenished aquifers.'

Restlessness trickled through the hall. Lois had said that in a supportive crowd a heckler would be howled down. If fakery or ego were discerned, outrage would run like wildfire, insults would fly fast. Speaking truly from the soul, where mind, heart and spirit flared, might not have worked.

'If we listen out for what the aquifers are saying, we'll all know what to do. They're receding fast. Seems to me they're saying, leave us alone, let us heal. Thanks.'

Relief merged with exhaustion and the powerlessness of entreaty. Helen and Kim started the applause, which spread slowly at first, then like fireworks, crackling on and on. Frankie Pankhurst, once the source of derision, was now looked upon with admiration. Not as her mother, Claire had once cherished her, but with collective, nourishing respect. Frankie's words, she could only hope, had ignited ideas. Frankie thought for an instant of how a rainforest regenerates from fire, its ectodermic buds sprouting in shades of crimson and gold. She had spoken truly, had been well-informed and honest. She had won them over.

Councillor Sheridan took the mic and invited Kasha Duff to speak on behalf of his application. He strutted to the front, took the microphone and turned to the crowd, like a cockerel turns to an intruder, head lowered and feathers up.

'If you don't want progress, go home and die. It's all inevitable.'

'Who wins, Kasha?' Don called from the crowd.

'You sure won't. You can't afford to stop progress.' Kasha slammed the mic on the table, everyone jumped at the crack. He swaggered to his place at the back of the hall.

Councillor Sheridan started to sum up. The audience was shuffling chairs. Lois stood and with a piercing voice invited everyone to stay for afternoon tea. It worked. Frankie looked through the crowd to see Kasha walking out. Bernard bobbed, not knowing whether to stay for the cake or leave with his boss. Small glows of determination warmed the chests and cheeks of Cawling's hardest nuts. Over tea and scones, ham and egg sandwiches, Frankie had her back slapped, her hand shaken. She kept saying, it was not up to her, they were all in it together.

Frankie develops courage and skills for public speaking through her involvement with friends and colleagues in Save Our Groundwater. As Matthew Cole argues, locating 'protagonists within networks that foster reciprocal awareness and cooperative action' is a means to 'reconfigure the train of action' (2021, p. 15). Confronting dilemmas of water scarcity or climate change 'as isolated individuals leaves citizens powerless' (Cole 2021, p. 15).

Groundwater narratives, built on the coherence of human and nonhuman agency, aim to evoke socially responsible collaborations.

The chapter argues that for groundwater's voices to be represented in public and scholarly discourse, they are usefully mediated through imaginative groundwater narratives that engage with the five main themes of this book: considering the ways writers know about groundwater as a posthuman subject; elaborating storying *with* that respects engagements with diverse stakeholders; incorporating accurate hydrogeology and associated sciences with aesthetic creativity; engaging with slippery concepts of nonrepresentation in evoking the place of groundwater; and building affective narratives that engage the emotions of readers. In storying *with* groundwater through climate writing, groundwater narratives join the choir of voices making a claim for the moral imperative for humans to engage in participative engagements with nonhuman actants. Primarily this book contends that the concept of narrative groundwater enables storying *with* in order to immerse readers in the material connections between human and more-than-human entities.

Four interrelated conclusions emerge. First, narrative groundwater generates literary responses to the Anthropocene through fiction, poetics or creative non-fiction, writing forms and styles that coalesce subterranean imaginaries within environmental and political contexts. Groundwater narratives acknowledge cultural and political contexts of anthropogenic threats to aquifers. Attention to epistemic aspects of groundwater's actions sits alongside relational consideration of human entanglements with subterranean waters. Practising 'response-ability' in climate writing entails acknowledging the impacts of human interferences through economic and political systems that channel benefits to minority wealth holders. It includes an address to social, environmental and cultural injustices. Groundwater narratives are one type of climate change literature, orientated towards not only social justice and distributive equality, but also political ethics that respond to many aspects of contemporary impacts of climate change on nonhuman lives and entities (Amin and Thrift 2013, pp. 189–193). Writing groundwater narratives in the Anthropocene enables authors across literary genres to challenge the social, economic and political forces of extractive industries and so-called market-driven economies. Literature's capacity to foster political conviction should never be underestimated (Amin and

Thrift 2013). As a climate writer I take responsibility for the underpinning intent to challenge political complacency. Explorations of the political context into which climate writers write are unavoidable.

Second, groundwater narratives aim for knowledge making, and in doing so navigate the ageless quarrel between poetics and scientific rationality. This quarrel still holds ground (Webb and Brien 2011, p. 192). Establishing the confluence between hydrogeology and subterranean imaginaries is an osmotic-like process. Posthuman and non-representational viewpoints give epistemological and ontological depth to groundwater narratives. Groundwater narratives are comfortable with entanglements between creative writing processes, scientific and philosophical positionings.

Third, the book has suggested storying *with* as a process whereby authors can write to know a subject. Amidst the bounds and liberations of the logic of storytelling, groundwater narratives tell a story, and in doing so unearth narrative groundwater's voices. Groundwater narratives test out possible ways to give voice to this inanimate entity. Storying *with* subterranean imaginaries includes writing into the spaces of new enmeshments of science and climate writing, seeing the creative research as 'a site of radical experimentation' (Krauth 2011, p. 11). Storying *with* groundwater recognises that the 'generative capacity of creative arts research is derived from subjective, emergent and interdisciplinary approaches' (Bolt 2010, p. 3). Storying *with* groundwater in the Anthropocene means acknowledging that the outcomes of creative practice-led research are unpredictable and may take many forms.

Fourth, giving emphasis to affective forces and emotion aims to ensure conjunctions for readers between ideas, feelings and actions. As Ben Anderson (2016) highlights, affects occur through encounters. Diverse and uncertain affective responses are possible. Groundwater narratives have an important role to play in breaking traditional dualisms between the rational and the emotive. The excerpts from *Why we cry* presented in this book attempt to portray the significant effects of encounters between groundwater and human and nonhuman lives. Emphasising nonhuman affective capacity alongside human rationality and emotional responses to the Anthropocene is suggested as an avenue through which to forge environmental politics and actions. Writing/reading/rewriting for diverse affects and emotions becomes a strategy to counterbalance didacticism or proselytising through information dumping in climate writing.

Expressions of the wider affective connotations of groundwater, its domain, its consumption and its relationality, seep from its metaphoric potency. Groundwater narratives aim for groundwater itself to affect readers as their enmeshments with groundwater are revealed. An affective politics to the work of writing climate fiction is paramount.

The implications for new climate writing and scholarship that may emerge from groundwater narratives are manifold. The potential to apply the keystone themes from this book to other climate change topics arises. Numerous other nonhuman entities whose existence is challenged in the Anthropocene can be 'storied *with*'. Establishing matter's enmeshment with humans, invoking readers' emotions through metaphor, and merging posthuman epistemologies with the parameters of fiction, poetry or creative non-fiction, are tools for writing many Anthropocene scenarios: loss of forests, glaciers and icecaps; species extinctions and rescues from the brink; continued burning of fossil fuels for energy and transport alongside expanding sustainable energy sources; finding solutions to the plastics islands polluting the Pacific Ocean; mitigating ocean acidification and rising sea levels, to mention a few. By considering the epistemological foundations to storying *with* the enmeshments between humans and the more-than-human world, climate writers give themselves firm ground to imagine affective stories that herald hopes of positive change in the Anthropocene. I hope to inspire climate writers and readers to critically write and read more climate writing, to question extractive economies, and to think into the impacts of posthuman ontologies, particularly in relation to thinking *with* groundwater.

Storying processes that express human enmeshments with more-than-human entities increasingly engage with affective environmental politics. Groundwater narratives join art and literature that aims not for predetermined political solutions, but for stories that acknowledge an 'earthly politics', which understands the world as 'an assemblage of heterogeneous threads' (Amin and Thrift 2013, p. 73). Groundwater narratives contribute wide literary interventions that seek to hear and respond to the voices of nonhuman actants. An 'affective politics' that sustains 'the desire and impetus for an open, just and collectively rewarding world' underpins groundwater narratives (Amin and Thrift 2013, p. 158). The aesthetic and affective potency of subterranean imaginaries lays the groundwork for climate narratives that inspire hope that human and nonhuman actants can cohere in life-sustaining, rather than destructive, entanglements.

My aspiration is for readers/writers to consider storying *with* groundwater as a means to make groundwater more visible, therefore enabling humans to become more accountable for the many impacts groundwater extraction is having on groundwater sustainability. I remain committed to the potential of stories of hope. *Why we cry* loosely follows a heroine's quest story structure. Frankie Pankhurst, through hardship, courage and engagement with community, wins a localised battle in the war for environmental groundwater justice. As Latour says, 'fiction anticipates what we hope to observe soon' (2017, p. 255). Maintaining hope is a way to affirm human and nonhuman capacities to respond to the impetus for sustainable futures. Hope may not come easily in the face of slow violences caused by capitalist extractive economies, and it should always be distinguished from naive utopianism (Wiseman 2021). Climate writing and particularly groundwater narratives must 'wake the frog' to avoid the slow boil that follows lack of action on climate change (Rand 2014). Groundwater's eminence and emergence as a deep and evocative voice in politically effective and affective literature has many potentialities.

I write into a watery future, a time when groundwater flows are imperilled. Narrative groundwater is an important part of emerging water cultures, a vital element in the watery turn in literature and social consciousness. Readers can know and feel the prompts towards personal and political questions about humans' and more-than-humans' becomings through narrative groundwater's calls. Taking an 'inward migration' to dwell in the place of stories is a missive for First Nations and white setter writers alike (Wright 2022). Subterranean imaginaries that nourish an outpouring of new groundwater narratives are imperative as the global effects of climate change on aquifers escalate.

Notes

1 In addition to authors and titles mentioned, some contemporary Australian examples include eco-poets Frances A. Johnson, Evelyn Araluen and John Kinsella; eco-fiction writers Inga Simpson, Mirielle Juchau, Alice Robinson, James Bradley and Evie Wylde; regenerative agriculture writer Charles Massey; and climate writer Tim Flannery.

2 For a thorough examination of the tragedy of water abuse in the Murray–Darling Basin see Margaret Simons's 'Cry me a river' (2020). And for more

details on how the Australian water market has had devastating effects on the inland water supply see Hamilton and Kells (2021).

3 For more information on these organisations explore the websites for Seed Mob (n.d.), Australian Youth Climate Coalition (n.d.), School Strike 4 Climate Australia (2022) and Elders Action Network (n.d.), among a myriad of climate activist organisations in Australia and worldwide.

References

Amin, A. and Thrift, N., 2013. *The arts of the political: new openings for the left.* Durham, NC: Duke University Press.

Andersen, G., 2020. *Climate fiction and cultural analysis: a new perspective on life in the Anthropocene.* Abingdon: Routledge.

Anderson, B., 2016. *Encountering affects: capacities, apparatuses and conditions.* New York: Routledge.

Australian Youth Climate Coalition, n.d. *Australian Youth Climate Coalition* [online]. Available from: www.aycc.org.au/ [accessed 23 January 2023].

Barthes, R., 2001. The death of the author. *In:* V. Leitch, ed. *The Norton anthology of theory and criticism.* New York: W.W. Norton & Co., 1466–1470.

Baucom, I. and Omelsky, M., 2017. Knowledge in the age of climate change. *South Atlantic Quarterly*, 116, 1–18.

Bennett, J., 2010. *Vibrant matter: a political ecology of things.* Durham, NC: Duke University Press.

Bolt, B., 2010. Introduction: toward a 'new materialism' through the arts. In: E. Barrett and B. Bolt, eds. *Carnal knowledge: towards a 'new materialism' through the arts.* London: I.B. Taurus, 1–14.

Bradley, J., 2017. Writing on the precipice. *Sydney Review of Books*, 21 February [online]. Available from: http://sydneyreviewofbooks.com/writing-on-the-precipice-climate-change/ [accessed 22 March 2017].

Brodie, R., Lawrie, K. and Commander, P., 2012. Groundwater – lifeblood of the continent. *In:* R. Blewett, ed. *Shaping a nation: a geology of Australia.* Canberra: Geoscience Australia and ANU Press, 332–279.

Buell, L., 1998. Toxic discourse. *Critical Inquiry*, 24, 639–665.

Cameron, A., 2014. *The last pulse.* Sydney: Random House Australia.

Carson, R., 1962. *Silent spring.* Boston, MA: Houghton Mifflin.

Cole, M.B., 2021. 'At the heart of human politics': agency and responsibility in the contemporary climate novel. *Environmental Politics*, 31, 1–20.

Cresswell, I. and Murphy, H., 2016. *Australia state of the environment 2016: biodiversity.* Canberra: Australian Government Department of the Environment and Energy [online]. Available from: https://soe.environment.

gov.au/theme/biodiversity/topic/2016/freshwater-species-and-ecosystems [accessed 27 April 2018].

Eamus, D. and Froend, R., 2006. Groundwater-dependent ecosystems: the where, what and why of GDEs. *Australian Journal of Botany*, 54, 91–96.

Elders Action Network, n.d. *Elders Action Network* [online]. Available from: https://eldersaction.org/ [accessed 12 February 2023].

Felski, R., 2008. *The uses of literature.* Malden, MA: Blackwell Publishing.

Geoscience Australia, 2018. *Groundwater dependent ecosystems* [online]. Available from: www.ga.gov.au/scientific-topics/water/groundwater/understanding-groundwater-resources/groundwater-dependant-ecosystems [accessed 27 April 2018].

Gergis, J., 2022 *Humanity's moment: a climate scientist's case for hope.* Collingwood: Black Inc.

Ghosh, A., 2016. *The great derangement: climate change and the unthinkable.* Chicago: University of Chicago Press.

Gibert, J. and Culver, D.C., 2009. Assessing and conserving groundwater biodiversity: an introduction. *Freshwater Biology*, 54(4), 639–648.

Griebler, C. and Lueders, T., 2009. Microbial diversity in groundwater ecosystems. *Freshwater Biology*, 54, 644–677.

Hamilton, S. and Kells, S., 2021. *Sold down the river: how robber barons and Wall Street traders cornered Australia's water market.* Melbourne: Text Publishing.

Haraway, D., 2016. *Staying with the trouble: making kin in the Cthulucene.* Durham, NC: Duke University Press.

Hennessy, R., Cothren, A. and Matthews, A., 2022. Creating new climate stories: posthuman collaborative hope and optimism. *TEXT*, 26, 1–20.

Humphreys, W.F., 2008. Hydrogeology and groundwater ecology: does each inform the other? *Hydrogeology Journal: Special Edition – Hydrogeoecology*, 17, 5–21.

Krauth, N., 2011. Evolution of the exegesis: the radical trajectory of the creative writing doctorate in Australia. *TEXT*, 15 (1).

Latour, B., 2017. *Facing Gaia: eight lectures on the new climatic regime.* Cambridge: Polity.

Morton, A., 2022. Australia's new approach was a rare positive at COP27 – but now the need for action is all the more acute. *The Guardian*, 21 November.

Mulhall, S., 2009. *The wounded animal: J.M. Coetzee and the difficulty of reality in literature and philosophy.* Princeton, NJ: Princeton University Press.

Pain, S. 2005. Underground Australia. *New Scientist*, 6 August, 28–33.

Papertalk-Green, C. and Kinsella, J., 2018. *False claims of colonial thieves.* Broome, WA: Magabala Books.

Perina, G., *et al.*, 2018. Understanding subterranean variability: the first genus of Bathynellidae (Bathynellacea, crustacea) from Western Australia described through a morphological and multigene approach. *Invertebrate Systematics*, 32, 423–447.

Rand, T., 2014. *Waking the frog: solutions to our climate change paralysis.* Toronto, Ontario: ECW Press.

Schneider-Mayerson, M., 2018. The influence of climate fiction: an empirical survey of readers. *Environmental Humanities*, 10, 473–500.

Schneider-Mayerson, M., *et al.*, 2020. Environmental literature as persuasion: an experimental test of the effects of reading climate fiction. *Environmental Communication*, advance online publication. https://doi.org/10.1080/17524032.2020.1814377

School Strike 4 Climate Australia, 2022. *School Strike 4 Climate* [online]. Available from: www.schoolstrike4climate.com/ [accessed 23 January 2023].

Sedivy, J., 2017. Why doesn't ancient fiction talk about feelings? *Nautilus*, 20 April [online]. Available from: http://nautil.us/issue/47/consciousness/why-doesnt-ancient-fiction-talk-about-feelings [accessed 22 June 2017].

Seed Mob, n.d. *Seed* [online]. Available from: www.seedmob.org.au/ [accessed 23 January 2023].

Segal, M., 2017. The missing climate change narrative. *South Atlantic Quarterly*, 116, 121–128.

Simons, M., 2020. Cry me a river. *Quarterly Essay*, 77.

Swyngedouw, E., 2004. *Social power and the urbanization of water: flows of power.* New York: Oxford University Press.

Todd, Z., 2015. Indigenizing the Anthropocene. *In:* H. Davis and E. Turpin, eds. *Art in the Anthropocene: encounters among aesthetics, politics, environments and epistemologies.* London: Open Humanities Press, 241–254.

United Nations Climate Change, 2022. *COP27 reaches breakthrough agreement on new 'loss and damage' fund for vulnerable countries.* Press release, 20 November [online]. Available from: https://unfccc.int/news/cop27-reaches-breakthrough-agreement-on-new-loss-and-damage-fund-for-vulnerable-countries [accessed 19 January 2023].

Webb, J. and Brien, D.L., 2011. Addressing the 'ancient quarrel': creative writing as research. *In:* M. Biggs and H. Karlsson, eds. *The Routledge companion to research in the arts.* Milton Park: Routledge, 186–203.

Weik von Mossner, A., 2014. *Moving environments: affect, emotion, ecology, and film.* Waterloo, Ontario: Laurier University Press.

Wiseman, J., 2021. *Hope and courage in the climate crisis: wisdom and action in the long emergency.* Cham, Switzerland: Springer International Publishing.

Wright, A., 2022. The inward migration in apocalyptic times. *Meanjin*, 81, 57–65.

INDEX

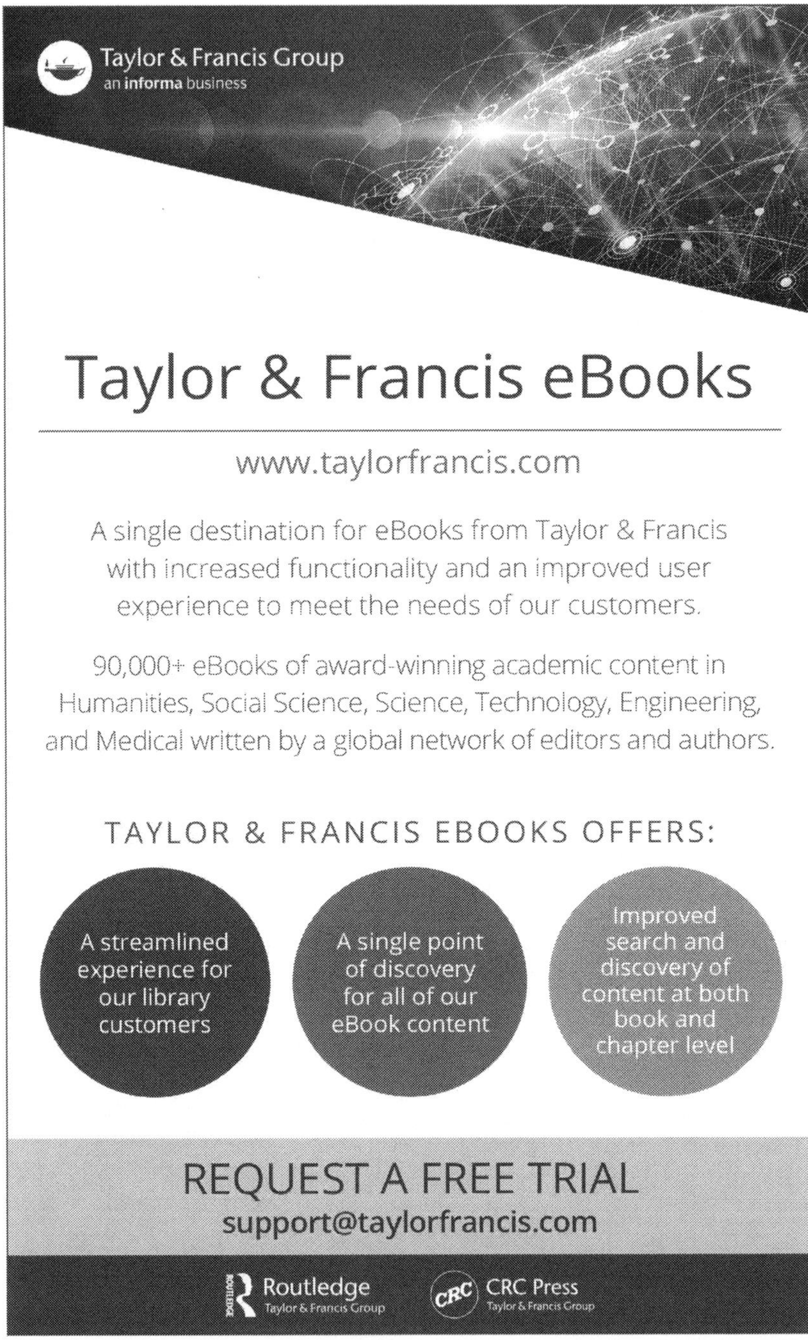

Taylor & Francis Group
an informa business
Taylor & Francis eBooks
www.taylorfrancis.com
A single destination for eBooks from Taylor & Francis with increased functionality and an improved user experience to meet the needs of our customers.
90,000+ eBooks of award-winning academic content in Humanities, Social Science, Science, Technology, Engineering, and Medical written by a global network of editors and authors.
TAYLOR & FRANCIS EBOOKS OFFERS:
A streamlined experience for our library customers
A single point of discovery for all of our eBook content
Improved search and discovery of content at both book and chapter level
REQUEST A FREE TRIAL
support@taylorfrancis.com
Routledge
Taylor & Francis Group
CRC Press
Taylor & Francis Group

Made in the USA
Las Vegas, NV
19 September 2026

fe1b6f78-6471-4b77-a910-ffeb2bb87952R02